Water Treatment Plant Operation and Management

Operational Guide to AWWA Standard G100

G100

Sarah C. Clark, PE

American Water Works Association

The Authoritative Resource on Safe Water®

Advocacy
Communications
Conferences
Education and Training
Science and Technology
Sections

Operational Guide to AWWA Standard G100
Water Treatment Plant Operation and Management

Disclaimer
The authors, contributors, editors, and publisher do not assume responsibility for the validity of the content or any consequences of its use. In no event will AWWA be liable for direct, indirect, special, incidental, or consequential damages arising out of the use of information presented in this book. In particular, AWWA will not be responsible for any costs, including, but not limited to, those incurred as a result of lost revenue. In no event shall AWWA's liability exceed the amount paid for the purchase of this book.

AWWA Publications Manager: Gay Porter De Nileon
Production: Glacier Publishing Services, Inc.

Library of Congress Cataloging-in-Publication Data
Clark, Sarah C.
Operational guide to AWWA standard G100 : water treatment plant operation and management / by Sarah C. Clark.
p. cm.
Includes bibliographical references.
ISBN 978-1-58321-853-2
1. Water treatment plants--Management--Standards--United States--Handbooks, manuals, etc. I. American Water Works Association. II. Title. III. Title: Water treatment plant operation and management.
TD433.C54 2011
628.1'6--dc23

2011022897

Printed in the United States of America
American Water Works Association
6666 West Quincy Ave.
Denver, CO 80235

Contents

American Water Works Association

G100 Operational Guide

Water Treatment Plant Operation and Management

SECTION 1: ACKNOWLEDGMENTS

Writing this guidance document has been a cooperative effort that involves the contributions of many people and utilities who are willing to share their knowledge with the water industry. The Water Treatment Plant Standard Committee members contributed significant time just to develop the approach to the standard and subsequently to write and refine the language. I am grateful to have had the experience in working with a terrific group of people to create the standard itself.

This operational guide is based on a format that was originally developed by James Ginley and Todd Humphrey, who created the first operational guide for Standard G400, Utility Management System, with input from Kanwal Oberoi. Oberoi provided the questions for the checklist at the end of this guide.

The following people kindly provided information and examples for this document:

Christine Owen, Tampa Bay Water, Tampa, Fla.
Jeff Boles, Town of Estes Park, Colo.
Mehrdad Morabbi, Austin Water, Austin, Texas
Charles Maddox, Austin Water, Austin, Texas
Kanwal Oberoi, Charleston Water System, Charleston, S.C.
Greg Moore, City of Golden, Colo.
Steve Hubbs, University of Louisville, Louisville, Ky.
Reid Campbell, Halifax Water, Halifax, N.S.

SECTION 2: FOREWORD

This operational guide has been written as a guidance tool for the implementation of water treatment plant operation and management best practices, as set forth in ANSI/AWWA Standard G100, Water Treatment Plant Operation and Management (AWWA G100; 2011b). AWWA G100 is part of the voluntary management standards for water utilities, initially established in 2004.

The original idea for the operational guide series to accompany the utility management standards came from utility managers who participated in a two-year pilot project that was conducted and supported by the American Water Works Association (AWWA) Utility Quality Programs Committee, the AWWA Accreditation Committee, and nearly 30 utility professionals from more than 10 North American utilities. The pilot project assessed the applicability and practicality of the series of AWWA utility management standards by working with two utilities—Birmingham (Ala.) Water Works and Sewer Board and Washington County (Va.) Service Authority—and a team of volunteers from utilities, consulting firms, and other organizations.

During the pilot process, the utility managers requested that AWWA develop some type of guidance or how-to documents to go along with the management standards. These guidance documents would serve two purposes: one, help utility managers understand the purpose and function of these new standards, and two, help them implement and incorporate the standards into everyday operations. From this suggestion the series of operational guides was conceived.

SECTION 3: INTRODUCTION

In order to consistently supply high-quality water drinking water to customers, the effective operation and management of drinking water treatment plants is essential. Critical requirements for the effective operation and management of drinking water plants are the subject of this guide, which provides suggestions and examples to assist water treatment plants in using AWWA G100.

The US Environmental Protection Agency's (USEPA's) Safe Drinking Water Act ensures public safety by setting extensive regulatory requirements for public water suppliers. However, progressive utilities may voluntarily subscribe to additional requirements, such as the AWWA utility management standards, in order to take a more proactive approach to guarantee that a high-quality product, which exceeds regulatory requirements, reaches the consumers' taps.

ANSI/AWWA Standard G100, Water Treatment Plant Operation and Management, is based on water treatment plant operation and management best practices, developed by consensus of the AWWA Standards Committee. AWWA G100 also calls for quality operation and maintenance of water treatment plants through the appropriate expertise of those operating and managing the system. This guidance document has been established to assist any water treatment plant, large or small, in adopting those practices set forth in AWWA G100.

This guide breaks AWWA G100 down into the following components:

Standard Language—The Standard Language is each section of AWWA G100, as written in the latest standard document.

Rationale—The Rationale gives background on the topic in each section of the standard and discusses why the requirements are important for water treatment plants.

Examples—The Examples given do not cover every aspect of the standard and will not apply to every utility's situation. They are not intended to be a complete instruction guide for the implementation of AWWA G100, rather they are intended to point interested parties in the right direction and give insight on how processes and procedures may be properly implemented in accordance with parts of the standard. In order to keep this publication to a reasonable size, only a handful of examples per topic were included, and most point to a secondary publication with current standard operating procedures and practices used by utilities in a standardized format. There are also many sources of additional information listed in the References and Resources section.

Questions to Check Progress—The Questions are listed as a tool to self-assess the status of a utility's compliance with AWWA G100. If a utility can confirm that it has all items in the questions in place, management may wish to consider applying for recognition, or the utility may be able to leverage that information when negotiating a bond rating.

The Audit Checklist is included to further assist utilities in performing an initial gap analysis or assessing their progress in implementing AWWA G100. This tool will help identify any gaps that may exist in current procedures and those required to meet the standard. It asks specific questions and looks for proof and documentation that procedures are indeed in place where called for by the standard. Just as the examples are not an instruction manual for implementing all of AWWA G100, neither is the checklist a blueprint. However, a utility that has all the components represented in the checklist is likely to have a majority of the AWWA G100 standard well covered in its operating and management procedures and practices.

SECTION 4: REQUIREMENTS

The minimum requirements for the operation of drinking water treatment plants can be divided into four major categories: regulatory compliance requirements, operational management practices, plant–real property management and maintenance, and water quality management. This standard requires that the utility establish measurable goals for each category, implement a monitoring system to measure performance against the goals, employ an optimization strategy to identify opportunities for improvement, and maintain consistent operation that satisfies the goals under conditions. Goals must be set with the concept of protecting public health foremost in mind, and shall include water quality and quantity targets.

Many relationships and linkages exist among the major elements of water treatment plant operation and maintenance. The requirements listed in this standard are meant to apply to water treatment plants. However, the system operator is cautioned to consider the effect of one requirement upon another based on site-specific conditions.

Sec. 4.1 Compliance With Regulatory Requirements

The plant shall satisfy the requirements of all applicable federal, state or provincial, and local regulations that apply to the operation of the plant. Plants shall demonstrate that as a minimum they meet the applicable drinking water regulations for their own jurisdiction.

Rationale

Customers expect their utility to meet or exceed all applicable regulations. In fact, maintaining consumer confidence demands that the utility maintain compliance. Protection of public health requires that water treatment plants produce water that meets all local, state, provincial, federal, or other regulatory requirements. The list of requirements that apply may be different for different treatment plants as some requirements are tied to the type of source water or treatment processes utilized. Responsibility for tasks that maintain compliance with the regulations applying to treatment plants might fall to personnel outside the plant staff, so coordination within the utility is essential to ensure that compliance requirements are met and documented. Plants that are initiating efforts to meet AWWA G100 and require coordination across the utility to meet regulations would profit from parallel application of ANSI/AWWA Standard G400, Utility Management System, and its companion *Operational Guide to AWWA Standard G400* (Ginley and Humphrey 2009).

One approach to managing regulatory requirements is to simply make a list of them, keep it up to date, and identify where the regulation can be found (e.g., a URL on the Internet or a book of codes). To make this list even more useful, identify the person(s) or staff position(s) responsible for ensuring compliance.

Example of Methods or Procedures

Example 1

Example 4.1 is an example Regulatory and Other Requirements Form from the Charleston (S.C.) Water System that can be used by a plant to identify and maintain a list of the most up-to-date versions of all regulations and other requirements that apply to the plant.

Regulatory and Other Requirements Form

Purpose/Scope: This document is used to identify and record legal and other requirements to which the utility subscribes, which have been identified as being directly applicable to the utility's activities, products, and services. Completion of this document will aid overall management of the water quality.

Instructions: This document is to be updated as regulatory and other requirements are modified, added, or removed. List the title of the requirement and provide a brief description in the column titled General Subject Covered. Rules, regulations, and laws must be periodically reviewed for current version and updates; therefore, identify which part of the utility will perform this review in the column titled Responsibility for Current Version. Establish electronic links in the column titled Document Format/Location. This ensures that the most current requirements are available for reference. With hard-copy requirements, list the location(s) of the hard copy.

All utility departments are required to ensure that regulatory and other requirements are interpreted and applied to their operations. Those listed as "Responsible" are required to periodically ensure that the most recent legal and other requirements are identified and listed. NOTE: This form may be replicated on a computer or duplicated on a photocopier. The computer copy must look similar to this document and contain the same information.

Prepared By/Date:

Title of Rule, Regulation, or Law	General Subject Covered	Responsibility for Current Version	Document Format/ Location (if electronic, provide Web address)

Source: Kanwal Oberoi, Charleston Water System, Charleston, S.C.

Example 4.1 Regulatory and Other Requirements Form

Example 2

Table 4.1 is an example list of permits with ongoing obligations from Tampa Bay Water (Tampa, Fla.) that is maintained to keep track of permits and the frequency of inspection in reporting. All regulatory requirements can be tracked on a spreadsheet that includes appropriate categories for the permits required within a specific jurisdiction. The utility also maintains a list of permits that have close-out obligations and are generally project specific, such as construction permits.

Table 4.1 Table of Permits With Continued Obligations

Permit or Permit Component	Inspection Frequency	Report Frequency	Renewal Frequency
Environmental Resources			
Mitigation			N/A*
Stormwater	Every 24 months	Every 24 months	N/A
Comprehensive Annual Reservoir Report		Annual	N/A
National Pollution Discharge Elimination System	Annual	Monthly	Every 5 years
Public Water Systems			
Monthly Operations Reports	Monthly	Monthly	
Comprehensive Regional Water Quality Monitoring Program	N/A	Monthly/quarterly/ 3-year/7-year/9-year	N/A
Hillsborough County Industrial Wastewater Discharge Permit		Monthly/summary report per discharge	Every 2 years
Underground Storage Tanks	Yearly		
Aboveground Storage Tanks			
Air Pollution Permit			
Development and Mitigation Approval			N/A
Aboveground Chemical Storage Tanks			

*N/A = not applicable.
Source: Christine Owen, Tampa Bay Water, Tampa, Fla.

Questions to Check Progress

1. Does the utility maintain and update regularly a list of legal and regulatory requirements?
2. Has the utility communicated this information to all the stakeholders?
3. Has the utility established programs to ensure that regulatory requirements are met?
4. Has the utility established a system of monitoring and evaluation to ensure compliance with the legal and regulatory requirements?

Sec. 4.2 Operational Management Practices

4.2.1 **Business Practices**

Each plant shall have a mission statement or purpose that includes optimizing water quality and that is communicated to individuals participating in the treatment of raw water for potable use. The treatment plant shall have, as part of its management goals, the review frequency of its performance goals. The management system shall keep records regarding the accomplishment of the goals, have a periodic review, and take documented action if goals are not being achieved. The following requirements apply to the portion of the utility's programs that pertain to the water treatment plant. Plant personnel shall be involved in the accomplishment of the utility's programs.

4.2.1.1 Budget and financial management. The plant shall assess the cost of operation at least annually and provide input regarding estimated expenses to accomplish the requirements of this standard. Plant personnel shall provide recommendations for equipment purchase and other plant-specific expenditures. Actual expenses shall be evaluated against estimates at least quarterly. The plant shall demonstrate adequate funds to incorporate aspects of the standard.

4.2.1.2 Treatment plant capital improvement planning. The plant shall have a current forward-looking assessment of the need for capital improvements consistent with master or facility plans. Master or facility planning should occur on a regular basis.

4.2.1.3 Energy management program. The plant shall review and optimize electrical energy usage at regular intervals, including review of energy use trends and costs, and shall consider energy costs in the evaluation of new or replacement treatment system components.

Rationale

To consistently produce high-quality water, a water treatment plant must set goals, and the plant personnel must understand the goals and how to meet them. Goal setting is part of the planning process and includes financial plans and budgeting. When plant personnel are involved in goal setting, including the budget and recommendations for equipment replacement and capital projects, the staff is more likely to take ownership of the goals and the budget and work toward meeting them.

Progress toward the goals should be reviewed and documented regularly in a framework that allows all personnel to participate in reaching the goals. Annual budgeting with at least quarterly checking of expenditures against the budget provides control of spending and the ability to make changes in spending as needed to meet changing conditions. A budget can be developed as a single spreadsheet, or it can be a fairly complex and long budget document. The size and complexity of the utility often impacts the amount of information normally included in an annual budget.

Capital improvement planning is often combined with a plant evaluation to determine what improvements are necessary to meet future demands, regulatory changes, or to replace aging infrastructure. The plant capital improvement plan, along with other financial information, is typically used to develop utility rate structures that cover current and planned improvements.

Energy management can be a cost savings program because energy costs continue to rise faster than inflation. Optimizing energy usage in a treatment plant requires an energy/sustainability team that will champion energy conservation. Access to energy billing information and benchmarking data for major pieces of equipment allows the team to plot monthly energy consumption costs and demand, which helps reveal where energy is being used. An orderly approach to optimization includes asking "what if" for each piece of equipment:

- What if the piece of equipment is turned off?
- What if the equipment ran at a lower capacity?
- What if more efficient equipment could be found?

The implementation program for energy conservation will be different for each plant, but each program should include a look at renewable sources, pursuit of grants and rebates for funding, and recognition that the program is a continuous improvement project.

Examples of Methods or Procedures

Example 1

Example 4.2-1 is an example of a utility vision, mission, values, and goals statement from Austin Water in Austin, Texas. This particular document is formatted into a wall poster so that it can be displayed in all areas of the utility.

Austin Water Vision, Mission, Values

Vision
To be recognized as the best water utility in the nation

Mission
To provide safe, reliable and high-quality water services to our customers

Values
Integrity
We adhere to the highest standards of ethics and will be accountable for all our actions.

Innovative
We continuously work toward and create new ways to improve our work processes and customer services.

Collaborative
We work as an organization to share information and help one another to achieve our goals.

Safe
We work in a way to minimize risk to ourselves and others.

Green
We implement ways to reduce our impact on our environment.

Diverse
We solicit and respectfully consider different perspectives from individuals with different backgrounds, ideas, and viewpoints.

Inspired
We promote a workforce that is motivated, positive, optimistic, and enjoys the work they do.

Goals
Workforce Development
Promote a workforce that is efficient, effective, inspired and accountable

Customer Value
Strengthen customer value and stakeholder relationships

Environmental Sustainability
Protect the environment through sustainable practices

Public Health and Safety
Protect the public health and safety by providing high-quality water service

Financial Management
Maintain strong financial position to ensure improved cost structure and competitive rates

Asset Management
Optimize life cycle cost of existing and future assets

Source: Austin Water, Austin, Texas.

Example 4.2-1 Vision, Mission, Values, and Goals for Austin Water, Texas

Example 2

Example 4.2-2 shows an example of the table of contents from a large utility annual budget. This document is issued annually not only for internal use, but also is available to the public and to other utilities that purchase water from Tampa Bay Water. In a smaller utility, only a subset of this information would normally be included in an annual treatment plant budget document, and it may all be done on a spreadsheet.

Tampa Bay Water 2011 Annual Budget
Table of Contents

Example 4.2-2 Table of Contents for Annual Budget for Tampa Bay Water, Tampa, Fla. (continues)

Financial Plan
- Funds Structure Overview
  - Fund and Accounts Flowchart
- Reserves and Restricted Funds
- Projected Changes in Fund Equity Schedule
- Financial Planning and Trends
- Revenues
- Expenditures
- Sources and Uses of Funds—2009 Actual, 2010 Approved, 2011 Approved
- Sources and Uses of Funds—Budget Projections 2012 through 2016
- Explanation of Variances
- Capital Improvements Planning and Financing
- Capital Improvement Program Definition
- Impact of Capital Improvement Program on the 2011 Operating Budget
- Routine Capital Purchases

Annual Rate Setting and Member Costs
- Overview
- Rate Setting
- Uniform Rate Elements
  - Fixed Costs
  - Variable Costs
  - Rate for City of Tampa From Tampa Bypass Canal
- Other Budget Factors
- Uniform Water Rates—2011 Approved, 2012 Through 2016 Projections
- Projected Water Demand for Fiscal Year 2011—Chart
- Estimated Annual Member Costs Fiscal Year 2011—Detail
- Estimated Net Member Costs Fiscal Year 2011—Chart
- Estimated Member Costs—Hillsborough County
- Estimated Member Costs—City of Tampa
- Estimated Member Costs—Pasco County
- Estimated Member Costs—City of New Port Richey
- Estimated Member Costs—Pinellas County
- Estimated Member Costs—City of St. Petersburg

Debt Service
- Bonds
- Rate Covenant
- Facility Acquisition Credits
- Summary of 2011 Debt Service
- Rating Agency Analysis
- Future Borrowing

Example 4.2-2 Table of Contents for Annual Budget for Tampa Bay Water, Tampa, Fla. (continued)

Source: Christine Owen, Tampa Bay Water, Tampa, Fla.

Example 4.2-2 Table of Contents for Annual Budget for Tampa Bay Water, Tampa, Fla. (continued)

Example 3

A system water plant budget is shown in Table 4.2-1 for a water system serving about 12,000 people. This budget covers the treatment costs for the Town of Estes Park, Colo. Source water, distribution, and billing costs are separate items in this budget and are not shown here.

Table 4.2-1 Town of Estes Park Purification Budget

	Budget		Projected			
	2010	2011	2012	2013	2014	2015
Personnel	$291,638	$277,998	$290,091	$302,710	$315,878	$329,618
Insurance Premiums	1,933	1,366	1,380	1,393	1,407	1,421
Professional Fees/ Service–Lab Testing	20,900	22,480	22,705	22,932	23,161	23,393
Publishing	1,000	1,000	1,010	1,020	1,030	1,041
Repairs and Maintenance						
Building Repair and Maintenance	7,000	32,000	32,960	33,949	34,967	36,016
Machinery/ Tools	150	150	155	159	164	169
Laboratory	3,000	14,000	14,420	14,853	15,298	15,757
Pump Treatment	2,050	2,050	2,112	2,175	2,240	2,307
Storage Tank	8,500	13,000	8,755	9,018	9,288	9,567
Maintenance/ Sample Stations	17,000	17,000	17,510	18,035	18,576	19,134
Materials and Supplies						
Office Supplies	200	225	232	239	246	253
Postage	3,500	3,100	3,193	3,289	3,387	3,489
Printing/Forms	2,950	3,100	3,193	3,289	3,387	3,489
Janitorial Supplies	300	300	309	318	328	338
Small Tools	350	650	670	690	710	732
Small Hardware	2,800	2,850	2,936	3,024	3,114	3,208
Chemicals	150,000	150,000	154,500	159,135	163,909	168,826
Laboratory/ Purification	21,850	26,850	27,656	28,485	29,340	30,220
Utility Expenditures						
Telephone	4,866	4,900	5,096	5,249	5,406	5,569
Natural Gas	25,000	25,000	26,000	26,780	27,583	28,411
Electric	43,679	60,000	62,400	64,896	67,492	70,192
Sewer	37,500	37,500	39,000	40,170	41,375	42,616
Propane Gas	5,000	5,100	5,304	5,463	5,627	5,796
Total Purification	$651,166	$700,619	$721,584	$747,269	$773,916	$801,561

Source: Jeff Boles, Town of Estes Park, Colo.

Example 4

A portion of the plant capital improvement plan (CIP) for the Sheridan Water Treatment Plant in Sheridan, Wyo., is shown in Table 4.2-2. This plan was developed from a plant and regulatory compliance evaluation with the goal of ensuring compliance and high-quality water well into the future. Total cost for each year is normally shown at the bottom of each column so that those cost numbers can be used for financial planning. More detailed project descriptions may accompany this type of a tabular listing as part of a larger CIP development document.

Table 4.2-2 Portion of Sheridan Water Treatment Plant Capital Improvement Plan

Project No.	Project Name	Year Project Completed								
		2010	2011	2012	2013	2014	2015	2016	2017	2018
S–1	Influent chemical feed & mixing improvements				$600–$700K					
S–2	Floc/sed basin flow management improvements				$500–$600K					
S–3	Floc/sed basin settling improvements				$1,400–$1,500K					
S–4	Polymer dose distribution upgrade				$50–$60K					
S–5	Floc/sed building roof replacement				$200–$300K					
S–6	Chemical feeders replacement									$250–$300K
S–7	Filter media replacement			$200–$300K						
S–8	Backwash water supply improvements			$400–$500K						
S–9	Clearwell flow monitoring improvements			$200–$300K						
S–10	Chlorine feed system improvements			$200–$300K						
S–11	*Cryptosporidium* disinfection improvements						$4,000–$5,000K			
S–12	Treated water quality control upgrade			$100–$200K						

Source: City of Sheridan, Wyo.

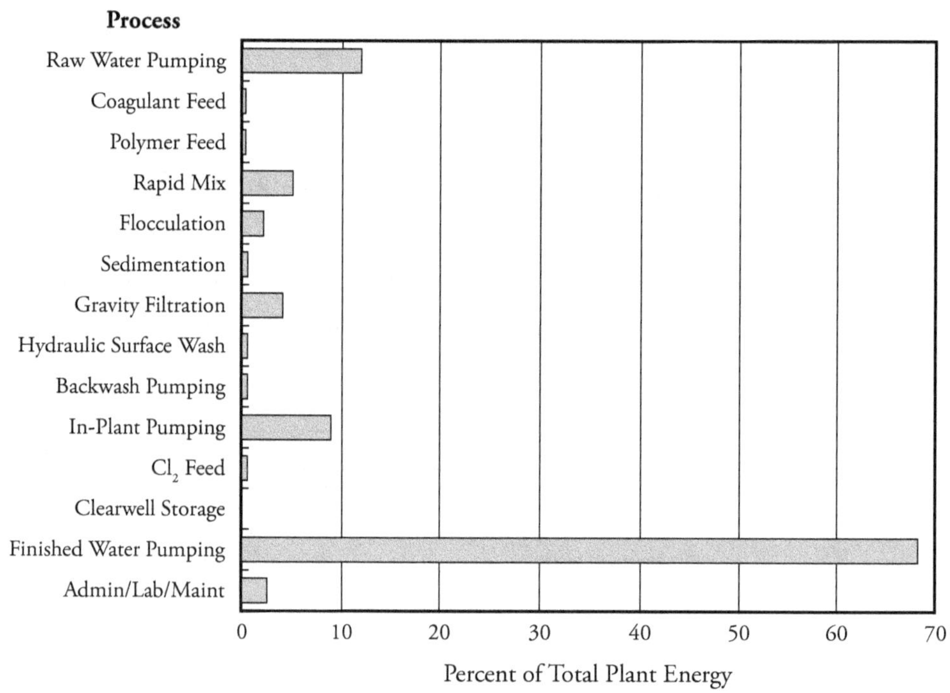

Figure 4.2-1 Typical Distribution of Energy Use in a Water Treatment Plant

Example 5

Figure 4.2-1 shows the distribution of and percentage of energy use in a typical water treatment plant. Pumping uses the major portion of energy in a plant, which suggests that initiating an energy optimization program by reducing power usage in pumping is an obvious place to begin saving energy and related costs.

Questions to Check Progress

1. Does the plant have a stated mission or purpose and performance goals that are updated regularly and communicated to the staff?
2. Does the plant develop a budget annually that is evaluated against expenditures each quarter?
3. Does the plant have a current capital improvement plan?
4. Does the plant have a systematic method for recording energy usage? Is energy consumption reviewed regularly and evaluated to determine whether reductions are feasible?

4.2.2 **Treatment Plant Production Limits**

The plant shall have a written policy that establishes the plant production limits for each unit process, sets process objectives, and compares the plant to permitted, approved, established production limits or design capacity appropriate for the raw water conditions.

Rationale

All water treatment processes are designed to operate within a set maximum flow rate to meet water quality targets set at the time of design. Production limits are important to ensure adequate treatment by all types of processes, including those that are more plant flow dependent (settling) and those that require adequate chemical dosing and mixing. Production capacity of unit processes is one of the measures used by the Partnership for Safe Water Phase III reviews. With a written document that establishes the production limits for each process, operators can easily check to make sure that limits are not exceeded and the plant protects against operating beyond the design limit and the attendant risk that water quality targets cannot be met.

Example of Methods or Procedures

Example 1

As is often done, Austin Water (Austin, Texas) incorporates process descriptions and capacities into its operation and maintenance (O&M) manual. These same descriptions could be part of a standard operating procedure (SOP), or they might be individual documents stored in a central location for reference by operators. In Austin, the O&M manual is in electronic form, but the same information could easily be included in a paper document. Example 4.2-3 is an extract from the Austin Water O&M manual which defines the process flow limits for rapid mixing and clarification at the Ullrich Water Treatment Plant.

Rapid Mixing, Flocculation, and Sedimentation

Description

There are seven up-flow solids contact reactor-type clarifiers at the Ullrich Water Treatment Plant (WTP). Basin size and capacity are shown below:

Basin Number	Size, ft	Q Rated, mgd	Q Max, mgd
1	150	35	37.5
2	150	35	37.5
3	110	20	22.5
4	110	20	22.5
5	110	20	22.5
6	110	20	22.5
7	110	20	22.5
8	110	20	22.5

Additional Design Criteria for Basin Nos. 1 and 2:

Design Treatment Flow Rate per Unit	35 mgd
Maximum Treatment Flow Rate per Unit	37.5 mgd
Maximum Hydraulic Flow Rate per Unit	45.0 mgd
Rise Rate at Design Flow	1.19 gpm/ft^2
Rise Rate at Maximum Treatment Flow	1.28 gpm/ft^2
Rise Rate at Maximum Hydraulic Flow	1.53 gpm/ft^2
Reaction Zone Detention Time at Design Flow	22.7 min
Maximum Recirculation Rate	187,400 gpm
Rake Drive Duty Rated Torque	400,000 ft-lb
Rate Mechanism Steel Design Torque	400,000 ft-lb

Source: Austin Water, Austin, Texas.

Example 4.2-3 Excerpt From Austin Water O&M Manual Defining Flow Limits for Rapid Mixing and Clarification

Example 2

Example 4.2-4 is a description of the filtration and backwashing processes that includes process objectives and turbidity limits from Austin Water's O&M manual. A simple daily comparison of these design limits and measured flow data allows the utility to ensure that operational flows are always within appropriate boundaries.

Filtration and Backwashing

Description

The purpose of filtration is the removal of particulate impurities and floc from water by passing it through a filter media for both aesthetic and public health reasons. The impurities typically consist of suspended particles (fine silts and clays), colloids, biological forms (bacteria, viruses, and plankton), and chemical precipitates from the lime-softening process. Filtration is both a physical and chemical process. Removal of turbidity in a filter is affected by the following factors: chemical characteristics of the water being treated, nature of suspension (physical and chemical characteristics and particulates suspended in the water), types and degree of pretreatment (coagulation, flocculation, and sedimentation), and filter type and operation.

The filters at the Ullrich Water Treatment Plant (WTP) are dual-media, high-rate, gravity-type filters. There are 18 filters arranged in two parallel rows. Each filter has a surface area of approximately 1232 ft^2. Each filter was designed to have 6 inches of sand, and 36 inches of anthracite. The coarsest material, the anthracite, also has the lowest specific gravity so that it remains the top layer. Each filter has a low profile Leopold Underdrain and integral media support cap. The underdrain system is designed to collect the filtered water evenly along the bottom of the filter and, similarly, distribute the backwash water uniformly. The maximum loading rate is 7.6 gpm/ft^2. Since all the valves are activated by the same control point, changes in flow are spread evenly over all the filters in service. As a result, hydraulic surges through the filters are minimized. Additional filters are brought into service as necessary to maintain the filtration rate at or below the target loading rate per filter of 13.4 mgd. The loading rate of 7.6 gpm/ft^2 allows the plant to achieve 175 mgd firm filtered water production capacity with 5 filters out-of-service (maximum loading of 13.4 mgd per filter). Dirty filters are never returned to service without first being backwashed.

When a filter is brought online, the filter produces water with a higher turbidity for several minutes. This period is called the ripening period of the filter run and generally lasts approximately 15 minutes. Peak filter ripening turbidities are negligible at Ullrich. The ripening algorithm executed by the Plant's Distributed Acquisition and Control System limits the ripening spike to a peak of approximately 0.12 ntu with a recovery back below 0.1 within 15 minutes or less. As a filter run proceeds, foreign material collects in the media, causing turbidities to drop and stabilize before rising again at the end of the filter run. A filter is backwashed to remove the accumulated material from the media after a designated time period. Before a filter is removed from service, another filter is slowly brought online to minimize flow changes and optimize filter performance.

The filter is taken out of service during the backwash process. The backwash system at the Ullrich WTP utilizes both air scour and water to clean the media and flush out the impurities and floc. Centrifugal pumps supply filtered water for backwashing. The backwash pumps are located at the High Service Pump Station. The blowers are located in the lower level of the Administration Building. One of the backwash pumps acts as a standby pump. The filters are backwashed using the high service pumps and water from the clearwell.

Example 4.2-4 Description of Filtration Process, Flow Limits, and Objectives (continues)

A functional description of the normal operation of one of the Ullrich WTP's filters follows:

1. During normal operation, settled water from the basins enters the filter through an inlet valve.
2. The water enters the filter gullet, flows over the washwater troughs (which are submerged during normal operation), and flows by gravity through the filter media.
3. The filtered water is collected in the underdrain system, conveyed to the filtered water collection conduit, and passes into the filter outlet piping in the pipe gallery.
4. The filtered water then passes through the outlet control valve into the filtered water pipe. This filtered water pipe collects the outlet flow from all filters and flows to the finished water clearwells.

Filters are backwashed automatically after a specified run time. For a general description of the backwash process, see the Process Assistance System. A functional description of the backwash process follows:

1. Washwater is supplied from the backwash pumps through the backwash header, flows through the backwash inlet valve, into the filtered water collection conduit, into the filter underdrain system, and up through the media.
2. The dirty washwater rises up into the washwater troughs, drops into the filter gullet, and flows out through the backwash outlet valve into the waste washwater conduit. The waste washwater is conveyed form the conduit to the waste backwash water pipeline, which flows by gravity to the washwater recycle basin. After backwashing, the filter may be returned to service.

Source: Austin Water, Austin, Texas.

Example 4.2-4 Description of Filtration Process, Flow Limits, and Objectives (continued)

Questions to Check Progress

1. Does the plant have defined production limits that are documented?
2. Does the plant have written descriptions of how each treatment process works and are these easily accessible to operators?

4.2.3 Treatment Plant Adequacy of Treated Supply

Plant production capacity should be adequate to meet system demands, taking into account the availability of system storage. The plant shall identify and meet system demands, have a plan to meet those demands, and have a process to implement the plan. The process shall include a method for reviewing the implementation progress for adequacy.

Rationale

Periodic evaluation of system demands and projection of demand increases for a reasonable planning period into the future allows a water system to ensure that sufficient production capacity is available to meet customer needs. A demand projection is typically used to plan for plant expansions and to set up capital improvement plans and rate structures that will accommodate improvements required to meet expected increased demands. Once system demands are determined and a plan is in place to meet the demands, the water plant or system must have a process in place to implement the plan. The plan and its implementation should be reviewed regularly to ensure that timing and progress are adequate to have treatment capacity in place to meet demands.

Documentation of a demand projection is generally provided in a demand study or evaluation, which can be completed by the utility or plant or by contract. A demand evaluation typically includes an analysis of the service population and growth rate, an evaluation of current demands and usage rates, an evaluation of land use and build-out conditions, and a projection of water demand into the future, which is tied back to each of these parameters. Demand projections always contain some uncertainty as there are a number of assumptions made in the process of completing the analysis.

Example of Methods or Procedures

Potable water demand projection can be illustrated graphically. Figure 4.2-2 is an example of a projection for a Colorado utility that includes projected dates for plant expansion and replacement to meet the build-out demand for the system.

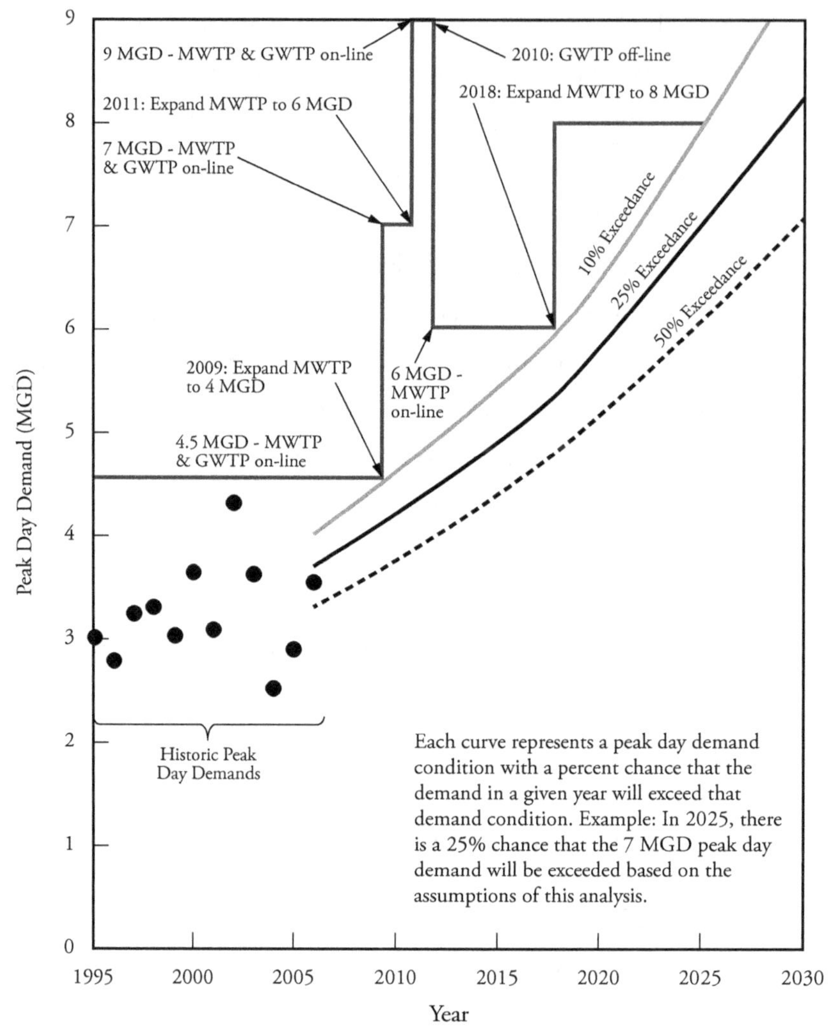

NOTE: MWTP = Marys Lake Water Treatment Plant; GWTP = Glacier Creek Water Treatment Plant.

Source: Town of Estes Park, Colo.

Figure 4.2-2 Typical Potable Water Demand Projection With Plan for Expansion

Questions to Check Progress

1. Has the plant identified current system demands and projected future demands?
2. Is there a plan in place for making required changes to meet future demands?
3. Is a process in place for implementing the plan and checking up on the progress of implementation?

4.2.4 **Housekeeping Procedures and Cleanliness**

Areas of the plant shall be maintained in a neat and orderly condition. Operation and maintenance functions should be conducted in such a manner that plant site safety and water quality are not compromised.

Rationale

Drinking water plants are producing water for public consumption that must meet regulatory health standards, so water plant facilities should be kept neat and clean to prevent accidental contamination of the treated water. In the same fashion that operators develop a commitment to making high-quality water, the plant staff should be encouraged to take ownership in the way the plant looks so that housekeeping becomes a habit, not a chore.

Orderly facilities contribute to efficient maintenance of equipment by allowing easy access to make repairs and a work environment that is conducive to getting the work done. Housekeeping procedures are closely related to safety in the prevention of accidents on the plant site. Particularly in plants that do not have excess free space, the work areas for making repairs must be clean and free of clutter to allow optimal working conditions for maintenance staff.

Example of Methods or Procedures

Figure 4.2-3 illustrates a treatment plant in Estes Park (Colo.) that has minimal spare space for working on equipment. Anything left in the walkways creates tripping hazards, and in some areas may compromise the functionality of the jib crane, jeopardizing the ability to remove and repair membrane cassettes. Housekeeping in this sort of environment is not only a safety necessity but also essential for an operator's ability to work on individual pieces of equipment without interference.

Source: Jeff Boles, Town of Estes Park, Colo.

Figure 4.2-3 Submerged Membrane Tanks With Jib Crane Tracks and Control Cabinets at the Marys Lake Water Treatment Plant

4.2.5 **Operating Permits and Licenses**

The plant shall obtain necessary operating permits and licenses in accordance with regulatory requirements. The record-keeping system shall be used to maintain current documents.

Rationale

All water treatment plants meeting this standard should be in compliance with applicable regulatory requirements. This includes obtaining and updating permits and licenses as necessary to meet the regulations. Utilities have many ways to track their regulatory permitting obligations, but most of them are simplified to a tabular or spreadsheet framework that is assigned to a particular staff member for updating. This information must be easily available to the groups responsible for reporting and renewal to prevent mishaps such as missing deadlines for renewal.

Example of Methods or Procedures

Example 1

Table 4.2-3 lists required permits and identifies the group or person within the utility who is responsible for inspections, reporting, compliance, and renewal of the permit. Organizing the requirements in this manner facilitates the process of ensuring that all reports and permits are kept up to date.

Table 4.2-3 Permits With Continued Obligations

Permit or Permit Component	Section Responsible for Inspection	Report	Renewal
Environmental Resources			
Mitigation	Evaluations & Permits	Evaluations & Permits	N/A*
Stormwater	Planning & Projects	Planning & Projects	N/A
Comprehensive Annual Reservoir Report	Source Rotation	Source Rotation	N/A
National Pollution Discharge Elimination System	Operations	Operations	Planning & Projects
Public Water Systems			
Monthly Operations Reports	Operations	Operations	Operations
Comprehensive Regional Water Quality Monitoring Program	N/A	Laboratory	N/A
Hillsborough County Industrial Wastewater Discharge Permit	Operations	Operations	Planning & Projects
Underground Storage Tanks	Yearly		
Aboveground Storage Tanks		Operations	
Air Pollution Permit		Operations	
Development and Mitigation Approval		Evaluations & Permits	N/A
Aboveground Chemical Storage Tanks		Operations	

*N/A = not applicable.
Source: Christine Owen, Tampa Bay Water, Tampa, Fla.

4.2.6 **Plant Management**

Management practices and policies shall support the accomplishment of requirements of this standard.

Rationale

Water treatment plants expecting to meet AWWA G100 must have support from plant management as demonstrated through daily management practices. A wide range of management issues play into successful implementation of the standard, including provision of sufficient and appropriately trained staff, ensuring that treatment equipment and facilities are adequate and maintained in good operating condition, and development and implementation of financial plans that allow for appropriate improvements.

A major national program often used to evaluate treatment plants is the Comprehensive Performance Evaluation (CPE) developed by the USEPA. This program has been used in all parts of the country for some time, and summary results from many of the evaluations show that management factors are the most frequent issues that impact water treatment plant performance. To commit to optimized performance, administrators must be willing to pursue actions aimed at improving plant performance. To avoid even momentary lapses in treatment that can increase the public health risk, an understanding of the public health implications of operating a water treatment plant should guide the management approach. Managers need to guard against complacency and develop an environment that encourages an empowered staff that can respond to all kinds of water conditions.

Refer to *Operational Guide to AWWA Standard G400, Utility Management System* (Ginley and Humphrey 2009), for a wealth of ideas regarding management improvement.

Questions to Check Progress

1. Is the plant maintained in a neat and orderly fashion?
2. Does the plant hold all required permits for operations, and are all permits tracked and updated as required?
3. Do management practices support the activities required to meet this standard?

4.2.7 **Quality Assurance Practices**

The plant shall have written operational practices and procedures for the operation and maintenance of the plant.

4.2.7.1 Record keeping and data review. Refer to ANSI/AWWA G400, Utility Management System (AWWA 2009b), for details on record keeping and data review.

4.2.7.2 Record review. A review cycle shall be established for operation and maintenance of the plant to verify that the limits of the operational control policy are satisfied, including, for example, water quality, laboratory operations, and mechanical equipment.

4.2.7.3 Regulatory reporting. The plant shall have a protocol to comply with the reporting requirements of applicable regulations. The protocol shall designate who is responsible for the reports and the timing of submittal. The protocol shall include enough lead time to acquire necessary approvals and still satisfy the deadline for submitting the reports.

4.2.7.4 Standard operating procedures. The plant shall have a system to develop and update standard operating procedures (SOPs) for equipment and plant production processes, and a system to develop and update operation and maintenance manuals. These procedures may be developed specifically for the plant or may be taken from manufacturers' literature or other appropriate sources.

Rationale

Most water treatment plants regularly collect significant amounts of operating and process control data that is used to operate the plant on a daily basis. This data, over a historical timeframe, can also be used to evaluate progress toward meeting goals and to identify weaknesses that would benefit from setting new goals and implementing activities to reach them. A program for analyzing and reviewing the data with respect to plant operations allows plant staff to verify progress toward operational goals and ensure that operating and maintenance procedures are met.

Water treatment plants are typically required to file regular monthly reports to a regulatory agency. Data for regulatory reports are generally collected daily and must be accumulated for inclusion in monthly submittals. Often this is accomplished electronically, but some plants still collect daily data on paper. Each plant should have a written standardized approach to gathering the appropriate information and completing regulatory reports.

For smaller plants where the staff functions for operations and data collection are centrally located and interdepartmental staff can easily communicate, the organization required for ensuring that monthly reports are completed, reviewed, and signed by the appropriate person and submitted to the regulatory agency on time may not require significant preplanning or organization. However, for larger utilities where the functions for plant operations may be completely separate from laboratory functions, the collection of regulatory reporting data may require a standard procedure to ensure that all the correct data is collected and combined into monthly reports for regulators. In either case, all water treatment plants should have a designated person who is responsible for ensuring that the data are entered into reports appropriately and that reports are delivered on time.

SOPs bring consistency to plant operations when operating teams change regularly and have differing views on how a water plant should be operated. Most water treatment processes are more efficient and produce higher quality water with consistent operations, as long as the process has been optimized to treat the raw water quality entering the plant. When raw water quality changes, a standard approach to deal with that change provides multiple operating teams with a similar understanding of the necessary treatment shifts. By extension, the finished water quality is generally more consistent.

Example of Methods or Procedures

Example 1

Development of SOPs in a water plant should be accomplished by a team effort that involves the appropriate staff members who have an understanding of the particular subject of the SOP. This is true of both operational and maintenance SOPs, because no single person knows everything that is needed for a specific procedure. Discussion and agreement about the content of an SOP brings a sense of team effort and ownership to the process that is invaluable in the long-term operation of a plant. Example 4.2-5 is a standard template for development of an SOP.

Example 2

Example 4.2-6 is part of an SOP for the filters at a treatment plant in Austin, Texas. This particular plant is controlled by a supervisory control and data acquisition (SCADA) system, so regular filter operations are controlled according to the control loop descriptions in the SCADA system. However, every operator must understand how to operate the plant manually, without electronic controls, because events can occur that require manual operation to keep the plant online.

Standard Operating Instruction Form

Purpose/Scope: This form is used to prepare Standard Operating Procedures (SOPs) for use by associates who perform operational control and maintenance management tasks. The department head (or designee) is responsible for the approval, revision, and issuance of SOPs and ensuring that associates have the necessary training to perform the job.

Instructions: Complete each section below prior to placing new or modified monitoring or control equipment in service. The department head (or designee) shall ensure that maintenance documents are available and controlled at all appropriate locations. NOTE: This form may be replicated on a computer or duplicated on a photocopier. The computer copy must look similar to this document and contain the same information.

Prepared by: Approved by/Date:

Title:

1. Purpose:

2. Scope:

3. Responsibility:

4. Work Preparation:

4.1 Labor:	4.2 Equipment:	4.3 Materials:

5. Work Step(s):

6. Related Documents:

Source: Kanwal Oberoi, Charleston Water System, Charleston, S.C.

Example 4.2-5 Standard Operating Instruction Form

Filtration and Backwashing

Operation

Operation Procedures
Operation Modes
Control Description

Operation Procedures

Standard Operating Procedures
Process Considerations
Turbidity Monitoring
Manufacturer's Operations Do's and Don'ts for the Media and Underdrain
Manufacturer's Operations Do's and Don'ts for the Blower
Routine Operating Checklist

Standard Operating Procedures

Standard Operating Procedure: Taking a Filter Out-of-Service and Bringing a Filter Online.
Standard Operating Procedure: Backwashing a Filter

Process Considerations

In order for the filtration process to be optimal, the operator must ensure that the coagulation and sedimentation processes are also optimized. The operator should strive for turbidity levels in the filter influent to be less than 2 ntu.

All filters in operation should be loaded at the same rate. This ensures that no one filter will become overloaded. The filter control system automatically provides an equal flow split between the operating filters.

Filter effluent turbidity is monitored continuously at each filter and at the computer control room. For additional information see the Process Assistance System.

Turbidity Monitoring

The filters at Ullrich WTP typically produce an effluent with turbidities of less than 0.07 ntu. It is the goal of the Utility to maintain turbidities of less than 0.1 ntu 95 percent of the time.

Turbidity from each filter is continuously monitored by in-line turbidimeters. An analog signal from each of the turbidimeters is sent to its respective filter control console where that signal is compared with turbidity setpoint values corresponding to the call for backwash and high turbidity alarm. Turbidity values are also displayed at any Plant Distributed Acquisition and Control System Human Interface/Machine Interface. In the event of a failure in the automated turbidity monitoring of any filter, operations personnel must collect and analyze samples from the filter and record the data to meet regulatory requirements. For more information see the Process Assistance System.

Example 4.2-6 Filter operating Procedure for the Ullrich Water Treatment Plant (continues)

Manufacturer's Operations Do's and Don'ts for the Media and Underdrain System
Leopold includes a list of Operation Do's and Don'ts in the Manufacturer's O&M Manual to protect the underdrain and the media:

Don't

- Backwash the filters with dirty water
- Backwash the system without adequate flow measurement and flow control devices operational. It is important that the backwash rate be limited to a rate that will not generate a differential pressure across the underdrain greater than 5.2 psi. This rate is typically about 40 gpm/ft^2, which is significantly greater than the design backwash rate.
- Subject the underdrain to hydraulic surges (water hammer) caused by turning pumps on with the valves open or opening valves too quickly. Valves should be set to open and closed smoothly in a minimum of 60 seconds.
- Increase the water backwash rate to a high value before purging air from the underdrain and media.
- Fill any dry filter, at any time, from the top.
- Subject to filter, during filtration, to frequent major changes in flow rate.

Do

- Subject the filter to periodic disinfection to reduce biological or algal growth.
- Provide and maintain flow monitoring and control equipment for the water and air backwash systems.

Manufacturer's Operations Do's and Don'ts for the Blowers
The air scour blower system supplies air to scour/backwash the filter media. The blowers are centrifugal blowers manufactured by Continental. For more information consult the manufacturer's O&M manuals.

Don't start the blower motor more than 3 times in one hour. And those starts should be spaced equally apart with a minimum of 20 minutes between starts. More frequent starts may result in severe damage to the motors.

Do

- Rotate the machine shaft as indicated by the arrow on the outlet or discharge head.
- Check all valves in the system and appropriately set valves to prevent the machine from surge operation.

Routine Operating Checklist

- Once a day, check the operation of the mechanical equipment and instruments in the filter gallery.
- Once a week, check the operation of the motor-actuated filter valves.
- Check that the valves which are supposed to be open are fully opened and the valves that are supposed to be closed are fully closed.

Example 4.2-6 Filter operating Procedure for the Ullrich Water Treatment Plant (continued)

- Monthly, lower the water level to the top of the filter media and visually inspect the filter bed for the following:
  - –Unlevel surface (mounds),
  - –Cracks,
  - –Unwashed areas,
  - –Color change of media, and
  - –Mud ball formation.

Normal procedure will be to initiate an automatic backwash sequence at the filter control panel. While backwashing look for the following:

- Uniformity of flow into the backwash troughs, and
- Excessive amounts of media being carried out by backwash flow.

Control Description
Overview
The major components of the Filter System are (18) filters, (2) air scour blowers, and (2) backwash pumps.

Associated Loops
Settled Water Still Well Levels
Filter Influent Valve
Filter Effluent Valve
Filter Drain Valve
Filter Washwater (Backwash) Valve
Filter Air Scour Isolation Valve
Filter Headloss
Filter Level Switches
Filter Effluent Turbidity
Main Backwash Header Pressure

Source: Austin Water, Austin, Texas.

Example 4.2-6 Filter operating Procedure for the Ullrich Water Treatment Plant (continued)

Example 3

Table 4.2-4 is a control document for the list of SOPs for the Bennery Lake Water Supply Plant near Halifax, Nova Scotia. When a new SOP is developed for this plant, the document is added to this master record of all the SOPs for this plant.

Table 4.2-4 Registry of Standard Operating Procedures for the Bennery Lake Water Supply Plant

Halifax Water	Environmental Management System Manual	Page 1 of 2
	OPERATIONAL CONTROL ISO 14001: 4.4.6	HW-EMS-12-04 September 2010

Bennery Lake Water Supply Plant REGISTRY OF STANDARD OPERATING PROCEDURES (SOPs)	
SOP–01	Handling, Storage, Use of Sodium Hypochlorite
SOP–02	Handling, Storage, Use of Potassium Permanganate
SOP–03	Handling, Storage, Use of Hydrated Lime
SOP–04	Handling, Storage, Use of Carbon Dioxide
SOP–05-01	Content Data Sheet
SOP–07	Handling, Storage, Use of Sodium Hydroxide
SOP–08	Handling, Storage, Use of Polyphosphate
SOP–09	Handling, Storage, Use of Alum
SOP–11	Refilling Sodium Hypochlorite, Drum Changeover
SOP–12	Refilling Potassium Permanganate, Drum Changeover
SOP–13	Refilling Lime Bulk Storage
SOP–14	Refilling Carbon Dioxide Bulk Storage
SOP–15	Refilling Sodium Hydroxide Bulk Storage
SOP–17	Spill Response–Polymer
SOP–18	Refilling Alum Bulk Storage
SOP–19	Filter Backwashing
SOP–21	Refilling Polyphosphate Bulk Storage
SOP–23	Spill Response—Sodium Hypochlorite
SOP–24	Spill Response—Potassium Permanganate
SOP–25	Spill Response—Carbon Dioxide
SOP–26	Spill Response—Petroleum Product (Bennery Lake)
SOP–27	Spill Response—Polyphosphate
SOP–28	Spill Response—Sodium Hydroxide
SOP–29	Spill Response—Alum
SOP–30	Operator Continuous Monitoring (Rounds)
SOP–31	Reporting Disruptions in Water Quality
SOP–33	Operation of Diesel Generator in Case of Emergency
SOP–34	Setting a Grundfos Pump to Manual Control
SOP–35	Handling, Storage, Use of Polymer
SOP–36	Refilling Polymer Bulk Storage
SOP–37	Response: Threatened or Confirmed Contamination of Raw Water
SOP–38	Response: Threatened or Confirmed Contamination of Treated Water
SOP–39	Refilling Petroleum Storage
SOP–40	Spill Response—Petroleum (Land)
SOP–41	Procedure: Petroleum Product Inventory Control

Source: Reid Campbell, Halifax Water, Halifax, N.S.

4.2.7.5 Treatment chemicals quality assurance. Chemicals added to water in the treatment process shall meet NSF/ANSI 60 standards for use in potable water or be approved by the applicable regulatory agency.

Rationale

Chemicals used for treating drinking water have the potential for adding contaminants to the drinking water if they do not meet appropriate standards. To avoid the risk of contamination from treatment chemicals, plants should require that all chemical products used in the treatment process meet NSF/ANSI 60 standards, Drinking Water Treatment Chemicals—Health Effects. Normally, this requirement is set at the time of procurement in the specifications for bidding chemicals.

Example of Methods or Procedures

Example 1

Example 4.2-7 is Charleston, S.C., water system's standard procedure for procuring chemicals and checking them on delivery.

AWWA standards provide guidance into specifying chemicals. In addition, the utility may require NSF and Department of Health and Environmental Control certification where applicable for all chemicals that may come in contact with water during the entire treatment process. Food-grade lubricants are used where appropriate as certain lubricant manufacturers may not have pursued NSF certification.

Aside from specifying the properties of the chemical, utility may also evaluate:

- Delivery capabilities and packaging where applicable
- Whether a supplier has multiple sources and/or manufacturing facilities
- Vendor's inventory and emergency assignment of inventory during disasters
- Compliance with ISO 14001 program
- Secondary suppliers
- Safety record, security information on drivers
- Delivery times to ensure that a full staff is available when hazardous materials are being unloaded

Standard Operating Instructions for Chemical Specifications

Safety Requirements:
Follow all requirements of the safety manual and the applicable safety program or procedure. This may include but is not limited to Lock-Out Tag-Out, Confined Space Entry, Hazard Communications (Material Safety Data Sheets [MSDSs]), Process Safety Management, Excavation and Shoring, Respiratory Protection, Electrical Safety, or Lab Safety.

Example 4.2-7 Standard Procedure for Procuring Chemicals (continues)

Use personal protective equipment (PPE) predicated by the work environment and material handling requirements. This may include but is not limited to eye protection, face protection, hardhat, foot protection, welder's equipment, life preservers, respirators, gas masks, SCBA, gas detection equipment, seat belts, hearing protection, or traffic vest. (Indicate location of safety manual and MSDSs.)

Specification for (Chemical):
The (chemical) must meet or surpass current (applicable AWWA standard). The chemical must also meet chemical/physical specification as stated in the most recent version of DHEC State Primary Drinking Water Regulations and/or have NSF/ANSI standard approval for use in potable water.

Additional Specifications:
The chemical must be commercially available in the form of (named chemical).

Basis for Rejection of Chemical:
If the material or its transporting container does not meet the chemical, physical, or safety requirements of the standard and the purchaser's specifications. The chemical meeting the requirements of this standard shall not introduce levels of contaminants into the water that exceed maximum contaminant levels (MCLs) as set forth by the Safe Drinking Water Act and its amendments when the water has been properly treated with the chemical.

The chemical meeting the requirements of this standard shall contain no substances in quantities capable of producing deleterious or injurious effects on the health of those consuming water that has been treated properly with the chemical. The chemical must meet a specified composition.

Packing and Shipping:
The chemical shall be transported by an approved carrier in a tank truck clean and free from contaminants and meet the requirements of the AWWA standard and/or be NSF approved. An affidavit of compliance meeting the requirements of the AWWA standard and/or NSF approval will be required by the purchaser and must be supplied with the bid quote in order to qualify for consideration. Each invoice should display a statement indication that the chemical meets the AWWA standard and/or NSF approval. (Delivery times should also be specified.)

All associates shall be made aware of the consequences of deviating from the task. Failure to follow task instructions may result in additional training and/or disciplinary action at the supervisor's discretion. The positive significant aspect for following this instruction is the maintaining of water quality thus protecting the health of the public consuming the potable water. The negative significant aspect for not following this instruction is loss of water quality and endangering the health of the public consuming the potable water.

Chemical Receiving:
It is imperative that shipment manifests are checked for every delivery to ensure that the correct chemical is received and unloaded into the assigned storage tank. Use an electronic log to record delivery information. Note acceptable delivery times.

- When a chemical truck arrives at the front gate, the guard will contact the control room operator that a chemical shipment has arrived.
- The guard will then log in the truck (time, registration, manifest) and direct him to the Designated Chemical Receiving Parking Zone as signposted.

Example 4.2-7 Standard Procedure for Procuring Chemicals (continued)

- The sign post instructs the driver to carry his Chemical Shipment Manifest to the Control Room. This is also called a bill of lading and the Control Room Operator and driver will sign and date this document and retain a carbon copy.
- The Control Room Operator shall check the Chemical Shipment Manifest for the Chemical Identification and direct the truck driver to the proper unloading location. Enter chemical delivery shipment and arrival time in the Electronic Log Book.
- The Control Room Operator will then contact another operator to meet the truck at the correct unloading point with the manifest.
- Assigned Operator meeting the delivery truck shall also check the Chemical Shipment Manifest, verify that the correct signatures have been secured, and the delivery hose is connected to the proper unloading point; verify that the hose connection is tight and the correct valve sequence is in place.
- For liquid chemical deliveries at the chemical tank farm, the spill protection mat must be placed over the storm drain prior to unloading. This mat is stored in a locker just in front of the berm. Ask the driver to alert Control Room Operator when they leave the facility; this is to ensure an accurate headcount in case of an emergency.

Alum shipments received after hours:

- Driver to use intercom at front gate to gain admittance.
- Process Control Operator opens front and inner gate once identification and contents have been verified.
- Follow steps 8–19* as usual.

Bulk Liquid Chemical Receiving:
(Alum, Caustic, Fluoride, Phosphoric Acid, Coagulant Polymer, Sodium Chlorite)

- Obtain dip sample from top of truck in a sealed container (driver can take the sample).
- Take the sample to the control room. Wear appropriate safety PPE. Carefully remove the cap and check the sample for proper color, unusual smells, or contamination debris.
- Carefully pour the sample into a 250-mL graduated cylinder.
- Place a thermometer into the liquid and read the temperature of the liquid for use in temperature correction. Record this reading on the chemical quality control sheet. If product temperature is greater than 125°F, do not continue. Allow tanker content's temperature to drop below 125°F, and repeat procedure beginning at Step 7*.
- Record the following: shipment number, date and time on chemical quality control sheet—sign quality control (QC) sheet.
- Remove thermometer and slowly insert hydrometer into sample and read the degrees Baumé or specific gravity. Record on the chemical quality control sheet.
- Add specific gravity or Baumé and the temperature correction together to obtain corrected Baumé or specific gravity.
- Check the corrected reading with the acceptable range at the top of the sheet to ensure that the sample is within the range.
- If acceptable, tell the driver to unload and ensure that the correct tank is available to receive the delivery.
- If not acceptable, resample and repeat Steps 7–15*. If still unacceptable, refuse delivery, call Operations Engineer and Director.

Example 4.2-7 Standard Procedure for Procuring Chemicals (continued)

- Label and save the QC sample until the next delivery comes in (M/D/Y; initials; time).
- Check that all documents have been received, signed, and dated; this includes bill of lading, delivery ticket, truck tare weight, QC form. All this information needs to be given to administration promptly.

Bulk Gas Receiving:
(Chlorine)

- Follow Process Safety Management Operating Guidelines for chlorine delivery.
- Control Room Operator will monitor SCADA system for chlorine gas alarms.
- Check that all documents have been received, signed, and dated; this includes bill of lading, delivery ticket, truck tare weight, QC form—all this information needs to be given to administration promptly.
- Check that electronic logbook has all relevant information.

All associates shall be made aware of the consequences of deviating from the task. Failure to follow instructions may result in additional training and/or disciplinary action at the supervisor's discretion. The positive significant aspects of following these standard operating instructions are the prevention of hazardous chemical release, prevention of environmental contamination, and prevention of violation of the conditional major air permit. The negative significant aspects from not following this instruction would result in violation of the conditional major air permit, hazardous chemical release, and possible violent chemical reaction, endangerment to life and health, and/or environmental contamination.

*Steps are part of the original documentation only and not part of this example.

Source: Kanwal Oberoi, Charleston Water System, Charleston, S.C.

Example 4.2-7 Standard Procedure for Procuring Chemicals (continued)

Questions to Check Progress

1. Does the plant have written standard procedures for plant operations?
2. Does the plant regularly review operating and maintenance data against operational goals and limits?

4.2.7.6 Chemical quality. The plant shall have procedures to establish the quality of delivered chemicals.

Rationale

Water treatment plants use numerous chemicals and take delivery of new supplies of chemicals on a regular basis. Each plant should have a standard procedure for checking the quality of the chemicals on delivery to ensure that chemicals meet required specifications. The procedure normally includes verification of information that is transmitted with the chemical delivery (labeling, packaging, or accompanying documentation). In addition, the procedure should describe sampling and testing of chemicals at a specified frequency upon delivery. Every utility should have the prerogative to reject a delivery, so the chemical delivery procedure should include decision-making criteria for chemical acceptance or rejection.

Example of Methods or Procedures

Example 4.2-8 is a procedure for unloading and checking chemical quality upon delivery from Austin Water's Ullrich Water Treatment Plant (WTP). A sample of ferric sulfate is taken from each delivery load and checked in the plant laboratory before offloading from the delivery truck into the plant storage tank.

Ferric Sulfate Unloading

1. Check driver's delivery ticket for proper information (Ferric Sulfate, Ullrich WTP).
2. Notify operators which storage tank to freeze.
3. Gather sample bottle and specific gravity meter.
4. Driver will pull sample from tanker.
5. Check alignment on valves on 4th floor.
6. Check suction alignment on the third floor.
7. Do not fill a storage tank that is "ON LINE."
8. Gather PPE—face shield, rubber gloves, ear plugs.
9. Check operation of safety showers in vicinity of tanker.
10. Check specific gravity (after steps 1–3 are complete).
11. Specific gravity should be any reading greater than >1,400.
12. Check with the driver on his hookup.
13. Turn air on regulator.
14. Set air about 36–38 psi (confirm with driver).
15. Open valve at ferric sulfate loading station (road side next to emergency generator).
16. Unloading time is between 1.50–2.00 hours.

Source: Austin Water, Austin, Texas.

Example 4.2-8 Ferric Sulfate Unloading Procedure for the Ullrich WTP

4.2.7.7 Equipment and instrumentation calibration. The plant shall have an instrument calibration program that includes the frequency of calibration. Equipment and instrumentation used to monitor the treatment process must be calibrated using traceable or certified standards or according to regulatory agency-approved or manufacturer-approved standards. The plant shall have a program for regularly calibrating the chemical feed equipment and performing a comparison between the chemical usages based on the feed equipment with the physical inventory at least monthly.

Rationale

Calibration of analytical instruments used for process control or for regulatory analysis using approved methods and certified standards is generally required by regulations. Even without the regulatory pressure, treatment staff needs to have confidence that water quality measurements by analytical instruments are reliable. This helps avoid second-guessing decisions or results. Operating a plant without calibrated measurements would be like driving a car with very poor visibility—unlikely to produce optimal results.

As water treatment plants have become more complex, the number of analytical instruments used to monitor the treatment processes has increased. To keep up with the calibration status of all instruments, a plant should have a written calibration program that can easily be followed. Staff responsible for calibration should be able to refer to the written program to verify timing of calibration for each instrument. Calibration results should be recorded and retained for reference. Programs should include information or references to information regarding how calibrations are to be done for each type and model of instrument.

Example of Methods or Procedures

Example 4.2-9 is a summary of a calibration program approach that is used by the Charlotte, S.C., water system. In addition to this type of general instructions for calibration, a complete program will include a listing of the instruments to be calibrated, frequency of calibration, and responsible party for ensuring calibration.

Calibration can be performed internally (a certified lab that can check various meters and process standards), although certain instruments require vendor calibration checks. Documentation and scheduling may be dictated by an Environmental Management System program. A vendor calibration list and schedule should include the type of equipment, manufacturer, model number, serial number, location, calibration frequency, calibration vendor, last calibration date, and the next calibration date. Real-time charts allow operators to compare their grabs with instrument readings; once the discrepancy goes beyond a predetermined tolerance, then an investigation should be performed to check instrument, reagents, and sampling/analysis technique.

Source: Kanwal Oberoi, Charleston Water System, Charleston, S.C.

Example 4.2-9 Summary of Calibration Program for Charleston, S.C., Water System

4.2.8 **Environmental Impact Management**

The plant shall identify and document discharges and comply with all applicable regulatory discharge requirements.

4.2.8.1 Disposal of discharges from the plant. Management of discharges, including chlorinated water, overflows, residual solids, air emissions, and stormwater shall be in accordance with applicable federal, state or provincial, and local regulations.

Rationale

Most water treatment plants have discharge streams of some sort, both regular flows and potential emergency streams. Each flow stream or discharge is subject to regulation by several entities, so treatment plant staff is responsible for understanding how these streams are regulated and obtaining the appropriate permits for discharging each of them. The environmental impacts of discharges will vary, depending on the type of discharge, so handling each stream may require a different set of operating procedures and staff skills.

Some utilities have developed SOPs for dechlorinating discharges that contain chlorine, particularly where the overflow can reach a natural water where the chlorine will affect aquatic life. Having a procedure in hand prior to an emergency overflow allows a more rapid and appropriate response.

Management of residual solids can involve a broad range of testing and permits, depending on the disposal method and site. The paint filter test is commonly used to determine the water content in solids prior to disposal in a landfill. The concentration of radioactive material in settled solids or in concentrate from membrane or ion exchange systems can put some utilities in the position of managing a hazardous waste, along with all associated extra permitting and handling requirements.

Example of Methods or Procedures

Example 4.2-10 shows portions of a discharge permit application for the Seawater Desalination Facility at Tampa Bay Water. Because this type of permit is managed by the local regulatory agency, applications will vary somewhat in different geographic locations. Basic information regarding permanent or temporary discharges is a common requirement for discharge permits. Note that the permit form requires that all discharge flows be identified, not just the flows being permitted under this particular application.

Section B—Business Activity

Give a brief description of all operations at this facility including primary products or services (attach additional sheets if necessary):

The Seawater Desalination Facility is designed to produce 25 million gallons per day of potable drinking water using reverse osmosis (RO) membrane technology with marine water obtained from the TECO Big Bend Power Station condenser cooling water as the raw water supply. The Process schematic is shown in Attachment A*.

Domestic wastewater is generated from the operation facilities. There are 17 employees on site. The domestic wastewater is discharged to the Hillsborough County sanitary sewer system.

The RO concentrate and treated filter backwash/rinse water are discharged back into Tampa Bay through a separate National Pollution Discharge Elimination System permit issued by Florida Department of Environmental Protection.

During normal operations, RO membranes are periodically cleaned; it is expected that six chemical cleanings will be performed per year. Two different chemical cleanings might be performed using one or more of the following chemicals: citric acid, hydrochloric acid, caustic, Ultrasil, EDTA (ethylenediaminetetraacetic acid), sodium tripolyphosphate, sodium docecylbenzene sulonate, and sodium lauryl sulfate. The maximum volume cleaning for one RO skid would be approximately 93,000 gallons. Note that the chemical solution will be prepared with a second-pass permeate water, which is a very low total-dissolved-solids water. During the cleaning cycle, the first flush of spent cleaning solution and rinse water is discharged to the scavenger tank. This solution is pH neutralized and tested prior to disposal into the Hillsborough County wastewater system. Description of the cleaning is provided in Attachment B*.

In addition, the diatomaceous earth (DE) filter and the cartridge filters (CFs) will also be periodically cleaned; it is expected that two cleanings a year will be performed for each DE filter and CF. For these filters, the same chemical cleanings as the ones for the RO skids will be used. The total volume for cleaning the seven CFs would be approximately 42,000 gallons, and for the 18 DE filters the total volume would be 148,500 gallons. Description of the cleaning is provided in Attachment B.

If the plant needs to be down, it is intended to preserve the RO membranes with sodium bisulfate. The preservation solution will be prepared with second-pass permeate. The spent preservative solution may also be directed to the scavenger tank prior to disposal to the Hillsborough County wastewater system. Hillsborough County will be notified when such event will occur.

Note that for the RO membranes, DE filters, and CF, a single chemical cleaning may be performed if two cleanings are not required at a given time.

COMPANY INFORMATION
Hillsborough County Occupational License/Code: License #148361, Code 380.002
Number of employees: 17 permanent employees currently
Hours of production operation: 24 hours per day **Days of week:** 7
Hours of business operation: 7:30–5:00 **Days of week:** M–F
Number of employees per shift: Minimum 2

Example 4.2-10 Excerpts From Discharge Permit for the Seawater Desalination Facility, Tampa Bay Water (continues)

Section H—Wastewater Flow and Characteristics

The User shall conduct an initial pollutant screening and submit the results of sampling and analysis for the wastewater constituents listed in Sec. 5.4 of Ordinance 99-12, as amended and the high-strength surcharge parameters identified in the County's rate resolution. If the wastewater flow is less than 25,000 gpd, sample a minimum of two times within a two-week period. If the wastewater flow is greater than 25,000 gpd, sample a minimum of three times within a two-week period. Samples shall be representative of daily operation, obtained through proper sampling techniques. This section need not be completed when first submitting the permit application. Please refer to the Industrial User Instruction Sheet for additional information.

Total facility flow in gallons per day (gpd):
Average 11,925; Maximum 144,255

Individual process flows in gallons per day (gpd):

Processes	Regulated (R) Unregulated (U) Dilution (D)	Average Flow Rate, gpd	Maximum Flow Rate, gpd	Type of Discharge (batch/continuous/none)
Seawater Intake	R—by FDEP	42,400,000	49,450,000	None to Sanitary Sewer
Clarifiers	R—by FDEP	2,230,000	2,600,000	None to Sanitary Sewer
RO Concentrate	R—by FDEP	16,960,000	19,780,000	None to Sanitary Sewer
Finished Potable Water	R—by FDEP	25,000,000	29,000,000	None to Sanitary Sewer
Membrane Cleaning Solution	R—by HCWD	11,745	144,000	Batch Discharge
Sanitary Water	D	135	255	Batch Discharge
Cooling Water	N/A			

*Attachments are part of the original documentation only and not part of this example.

Source: Tampa Bay Water, Tampa, Fla.

Example 4.2-10 Excerpts From Discharge Permit for the Seawater Desalination Facility, Tampa Bay Water (continued)

Questions to Check Progress

1. Does the plant utilize a standard procedure to verify chemical quality?
2. Is equipment calibrated regularly?
3. Are all discharges from the plant managed in accordance with regulatory requirements?

4.2.9 Emergency Preparedness and Response

The plant shall have documentation that describes emergency plans and provides specific directions to plant personnel in the event of an emergency. The program shall satisfy applicable regulatory requirements.

Rationale

Every water plant should have a written emergency response plan because when an emergency occurs, the operating staff is typically dealing with myriad problems and there is little time to think about the big picture and attend to all the problematic details. Some regulatory agencies require water systems to have emergency response plans, and in that case, the plan should address the regulatory requirements at a minimum. The whole intent of thinking through the potential needs in an emergency situation is to make emergency management more effective and efficient, to reduce the stress to plant and utility staff, and to facilitate the continuous provision of safe drinking water to customers.

An emergency response plan should identify the typical types of emergencies that could occur for that water plant. The plan should detail the major actions that may be required in any emergency, so staff has a list of options during a major incident. Response plans should be updated regularly, using plant experience in coping with incidents and ensuring that contact information is up to date.

Response plans should include contact names and telephone numbers for

- Local cooperative emergency response contacts such as fire, police, utility cooperative response groups, and emergency management agencies;
- Sources of emergency water, either interconnecting systems or bottled water sources;
- Local health officials;
- Government agencies that may be either required or available to assist, such as the regulatory agency; and
- Communication outlets (television, radio, and Internet).

The plan should include a description of the management structure to be used in the event of an emergency. Some plants adopt the Incident Command model used by many fire and police departments, so that approach would be reflected in the response plan structure. The plan should identify key utility employees, their roles in an emergency, and their contact information.

AWWA has developed a new utility management standard, AWWA Standard G440, Emergency Preparedness Practices, which will be available in late 2011 (AWWA forthcoming). ANSI/AWWA G440 and AWWA Manual of Practice M19, *Emergency Planning for Water Utilities* (AWWA 2001), provide extensive guidance on emergency planning and preparedness activities.

Example of Methods or Procedures

Example 1

Emergency response procedures should cover a wide range of situations that may occur in a water plant. Example 4.2-11 is the introduction to the emergency procedures document for the Ullrich Water Treatment Plant in Austin, Texas.

Emergency Procedures

Introduction

Emergencies are unexpected and unpredictable and take many forms. No one can be fully prepared for everything that may happen, but some simple measures are helpful in any emergency. Some things to remember when faced with an emergency are as follows:

- Think of everyone's safety first.
- Use common sense.
- Be factual and unemotional with other employees.
- Act quickly and calmly.

Using these simple steps when an emergency occurs will help determine what to do, when to act, who to call, where to go. For most emergencies, some preparations can be made. That is why these procedures should be read and understood by you and your co-workers because nobody knows when an emergency will occur. Are you ready for an emergency?

Identification

As stated earlier, emergencies come in all shapes and sizes and can occur at any time. Some of these emergencies we may be able to control, some may take some help to control, and some may be completely out of control. To help in identifying an emergency, think of a tuba:

- **T**hink of everyone's safety.
- **U**se common sense.
- **Be** factual and unemotional.
- **A**ct quickly and calmly.

These are the first steps in identifying an emergency. Our reaction and how we respond to an emergency type situation will determine its course and our safety. Some situations that could possibly and quickly turn into an emergency are the following:

- Fire
- Severe weather
- Power failures
- Bomb threats
- Explosive devices
- Health emergencies
- Chemical spills

Source: Austin Water, Austin, Texas.

Example 4.2-11 Introduction to Emergency Response Procedures for the Ullrich Water Treatment Plant in Austin, Texas

Example 2

Emergencies can arise from a wide range of events, one of which is wildfire in a water system watershed. These events can cause drastic changes in water quality in a very short period of time along with other major impacts. Utilities with forested watersheds have begun to develop fire response plans similar to the one in Example 4.2-12.

Short-Term Plan for Wildfire Response

A short-term operational plan must take into account the high likelihood that the surface supply will not be acceptable as an unfiltered supply for intermittent and possibly prolonged periods of time. Specific people should be identified to take responsibility for particular actions and a listing of local resources developed that the utility may need to use to implement the plan. The following list of action items will address the immediate need to continue delivering water to customers if there is a fire.

- Shut off the intake as soon as possible after the fire starts and begin using well supply. While major quantities of fire debris are not likely to reach the intake until the fire is out and the first rain arrives, there may be impacts from fire retardants reaching the stream during the fire. While travel time from the intake to town is several hours, the prudent approach is to ensure that no contaminants enter the pipeline. If contaminated or debris-containing water enters the pipeline, the hydro plant will be impacted as will the raw water ponds.
- Determine who is in charge of the fire-fighting effort and maintain contact with that agency so as to obtain maximum information regarding the extent and progress of fire control.
- Identify a source and arrange for use of appropriate heavy equipment to clear debris from the low head dam and intake structure. Debris will collect during the first flush runoff event at the intake area and will need to be cleared. Emergency permission to operate equipment in the stream channel may be required to maintain the intake in usable condition.
- In the event that cleaning of the impoundment is necessary after a fire, care must be taken to avoid bending or damaging the fish screen with heavy equipment. Because the screen has very small openings, it will likely function more like a wall with respect to large debris flushed out after a fire. Silt and sand may accumulate inside the screen and require cleaning inside the fish screened area. In this instance, cleaning may need to be accomplished by hand.
- Install a rain gauge upstream of the intake at a distance that will provide an early warning signal to the utility to prepare for high turbidity water. This will allow proactive closing of the intake valves to protect the pipeline, the hydro plant, and the raw water tank water quality.
- Notify the State of changed conditions in the watershed. Follow up with any actions the State requires that arise from the changed conditions.
- Develop an emergency monitoring plan that will allow the utility to obtain daily information regarding raw water quality. The monitoring should include turbidity, total coliform, total organic carbon (TOC), pH, iron, and manganese. With that daily information, decisions can be made as to whether the water can be used in the potable system or not.
- Define water quality criteria that serve as a basis for electing to use or not use the surface water supply. The supply cannot be used for potable water if the turbidity exceeds the regulatory limit. Limits on other water quality parameters such as total coliform, *E. coli*, TOC, iron, and manganese may also be criteria for shutting down the intake.
- Set up a system for notifying the public of water conservation requirements and determine the conditions under which mandatory conservation would be necessary. If a fire occurs in the summer during high pumpage conditions, conservation or restricted use of the well supply may be warranted in order to be able to provide a long-term supply using only the well sources.
- Assign an individual from the utility staff to participate in the development of a Burned Area Emergency Rehabilitation (BAER) team report. The utility will need a voice in the process of developing post fire rehabilitation plans for the watershed. BAER reports are specific to each fire and generally include estimates of the potential magnitude of the increases in runoff and erosion, an assessment of the risks posed by the increases, and recommendations for mitigation treatment on National Forest lands.

Source: HDR Engineering.

Example 4.2-12 Major Short-Term Response Elements of a Wildfire Response Plan

Example 3

Another emergency planning concern is the release of toxic chemicals in a water plant. The Toxic Chemical Release Response Plan for the Charleston, S.C., Water System can be found in Appendix A.

4.2.10 Security Planning and Implementation

The plant shall have documentation that addresses security issues and describes, in detail, the response of plant personnel in the event of a security incident. Elements that address the protection of the plant, personnel, and water supply shall be included.

Rationale

Security issues at water treatment plants range from minor incidents of vandalism and facility intrusion to major threats to facilities and water quality from terrorist activities. For most plants, daily security issues are most likely to involve access control or breaches of locks and fences. Security plans are essential to inform all operating staff of standard procedures to follow in a security incident to protect both the facility and personnel on site. Regular review of procedures with staff is necessary to ensure that each one understands actions required in the event of a security incident.

In the event of a security incident, contact with the local law enforcement agency is generally the first response. Many utilities have multiple facilities that may be located in different law enforcement jurisdictions. Thus the security plan for a particular water plant or wellhouse must be tailored to the geographic location so that the correct enforcement agency is contacted. For example, the Town of Estes Park has two treatment plants located outside the town limits, within county jurisdiction. The security plan for these plants requires that security issues be reported directly to the county authorities because if the town police are called first, they will just call the county, delaying the response time by the time it takes to make a second telephone call.

Because the focus on security for facilities containing chemicals has become a national concern, many facilities follow the security notification requirements of the US Department of Homeland Security.

Example of Methods or Procedures

Example 1

Tables 4.2-5 through 4.2-9 demonstrate a Security and Terrorism Response Plan that incorporates the US Department of Homeland Security rating system for terrorist attack levels.

Table 4.2-5 Security Response Procedures for a Low Threat Level

Condition	Standard Procedures	
LOW (GREEN) **Low Risk of Terrorist Attacks** Signifies a low risk of terrorist attacks. Protective measures should focus on ongoing facility assessments and the development, testing, and implementation of emergency plans. In addition to Threat Level Green, there are four higher threat levels: Blue, Yellow, Orange, and Red. (Please refer to the other fact sheets for information on suggested steps to be taken during other threat conditions levels.)	**Detection**	• Monitor water quality at the source water, leaving the plant, and in distribution and storage systems. Establish baseline results. Review operational and analytical data to detect unusual variations. • Follow up on customer complaints concerning water quality and/or suspicious behavior on the facilities. • Confirm communication protocol with public health officials concerning potential waterborne illnesses.
	Preparedness	• Post emergency evacuation plans in an accessible, but secure, location near entrance for immediate access by law enforcement, fire response, and other first responders. • Inventory spare parts and on-hand chemicals. Check if sufficient. • Identify sensitive populations within the service area (e.g., hospitals, nursing homes, daycare centers, schools, etc.) for notification, as appropriate, in the event of a specific threat against the utility. • Back up critical files such as plans and drawings, as-builts, sampling results, billing, and other critical information. • Conduct appropriate background investigation of staff, contractors, operators, and others with access to the facility. • Prepare vulnerability assessments and revise to incorporate changes made (e.g., assets added/replaced or new countermeasures implemented). • Ensure that employees understand appropriate emergency notification procedures.
	Prevention	• Train staff in safety procedures, such as handling hazardous materials and maintaining and using a self-contained breathing apparatus. • Secure equipment such as vehicle and spare parts. • Monitor request for potentially sensitive information.
	Protection	• Check all chemical deliveries for driver identification and verification of load. • Maintain vigilance and be alert to suspicious activity. Inspect buildings in regular use for suspicious packages and evidence of unauthorized entry. Report any suspicious activity to appropriate authorities. • Prosecute intruders, trespassers, and those detained for tampering to the fullest extent possible under applicable laws. • Review request for tours and identify protocols for managing the tour. • Implement controls for construction activities at critical sites. • Maintain disinfectant residuals as required by regulations. • Implement best management practices for optimizing drinking water treatment.

Source: Kanwal Oberoi, Charleston Water System, Charleston, S.C.

Table 4.2-6 Security Response Procedures for a Guarded Threat Level

Condition	Standard Procedures (plus those at lower threat levels)	
GUARDED (BLUE) **General Risk of Terrorist Attacks** Signifies a guarded risk of terrorist attacks. Protective measures should focus on activating employee and public information plans, exercising communication channels with response teams and local agencies, and reviewing and exercising emergency plans.	**Detection**	• Test security alarms and systems for reliability.
	Preparedness	• Reaffirm communication and coordination protocols (embedded in the utility's emergency response plan) with local authorities such as police and fire departments, HAZMAT teams, hospitals, and other first responders. • Prepare and/or revise emergency response plans and associated communication protocols. Include appropriate local officials concerned with law enforcement, emergency response, and public health. • On a regular basis post employee reminders about events that constitute security violations and ensure that employees understand notification protocol in the event of a security breach. • Prepare draft press releases, public notices, and other communications for a variety of incidents. Route through appropriate channels of review to ensure that pieces are clear and consistent.
	Prevention	• Secure building, rooms, and storage areas not in regular use. Maintain a list of secured areas or facilities and monitor activity in these areas.
	Protection	• Control access to mission-critical facilities.

Source: Kanwal Oberoi, Charleston Water System, Charleston, S.C.

Table 4.2-7 Security Response Procedures for an Elevated Threat Level

Condition	Standard Procedures (plus those at lower threat levels)	
ELEVATED (YELLOW) **Significant Risk of Terrorist Attacks** Signifies an elevated risk of terrorist attacks. Protective measures should focus on increasing surveillance of critical facilities; coordinating response plans with allied utilities, response teams, and local agencies; and implementing emergency plans, as appropriate.	**Detection**	• To the extent possible, increase the frequency and extent of monitoring activities and review results against baseline. • Increase review of operational and analytical data (including customer complaints) with an eye toward detecting unusual variability (as an indicator of unexpected changes in the product.) Variations due to natural or routine operational variability should be considered first. • Increase surveillance activities in source and finished water areas.
	Preparedness	• Review and update emergency response procedures and communication protocols. • Establish unannounced security spot checks (e.g., verification of personal identification and door security) at access control points for critical facilities. • Increase frequency for posting employee reminders of the threat situation and about events that constitute security violations. • Ensure that employees understand notification protocol in the event of a security breach. • Conduct security audit of physical security assets, such as fencing and lights, and repair or replace missing/broken assets. Remove debris from along fence lines that could be stacked to facilitate scaling. • Maximize physical control of all equipment and render vehicle inoperable when not in use (e.g., lock steering wheels, secure keys, apply a chain and padlock on front-end loaders, etc.). • Review draft communications on potential incidents, and brief media relations personnel of potential for press contact and/or issuance of release. • Review and update list of sensitive populations within the service area, such as hospitals, nursing homes, daycare centers, schools, etc., for notification, as appropriate, in the event of a specific threat against the utility. • Contact neighboring water utilities to review coordinated response plans and mutual aid during emergencies. • Review whether critical replacement parts are available and accessible.
	Prevention	• Carefully review all facility tour requests before approving. If allowed, implement security measures to include list of names and request identification of each attendee prior to tour, prohibit backpacks/duffle bags and cameras, and identify parking restrictions. • On a daily basis, inspect the interior and exterior of buildings in regular use for suspicious activity or packages, signs of tampering, or indications of unauthorized entry. • Implement mailroom security procedures. Follow guidance provided by the United States Postal Service.
	Protection	• Verify the identity of all personnel entering the water utility. Mandate visible use of identification badges. Randomly check identification badges and cards of those on the premises. • At the discretion of the facility manager or security director, remove all vehicles and objects (e.g., trash containers) located near mission-critical facility security perimeters and other sensitive areas. • Verify the security of critical information systems (e.g., supervisory control and data acquisitions (SCADA) systems, Internet, email, etc.) and review safe computer and Internet access procedures with employees to prevent cyber intrusion. • Consider steps needed to control access to all areas under the jurisdiction of the weather utility.

Source: Kanwal Oberoi, Charleston Water System, Charleston, S.C.

Table 4.2-8 Security Response Procedures for a High Threat Level

Condition	Standard Procedures (plus those at lower threat levels)	
HIGH **(ORANGE)** **High Risk of Terrorist Attacks** Signifies a high risk of terrorist attacks. Protective measures should focus on limiting facility access to essential staff and contractors, and coordinating security efforts with local law enforcement officials and the armed forces, as appropriate.	**Detection**	• Increase the frequency and extent of monitoring activities. Review results against baseline. • Confirm that county and state health officials are on high alert and will inform water utilities of any potential waterborne illnesses. • If a neighborhood watch–type program is in place, notify the community and request increased awareness.
	Preparedness	• Confirm that emergency response and laboratory analytical support network are ready for deployment 24 hours per day, 7 days a week. • Reaffirm liaison with local police, intelligence, and security agencies to determine likelihood of an attack on the water utility personnel and facility and consider appropriate protective measures (e.g., road closing, extra surveillance, etc.). • Practice communications protocol with local authorities and others cited in the facility's emergency response plan. • Post frequent reminders for staff and contractors of the threat level, along with a reminder of what events constitute security violations. • Ensure that employees are fully aware of emergency response communication protocols and have access to contact information for relevant law enforcement, public health, environmental protection, and emergency response organizations. • Inspect and practice activation of available emergency interconnections with neighboring water agencies. • Have alternative water supply plan ready to implement (e.g., bottled water delivery).
	Prevention	• Discontinue tours and prohibit public access to all operational facilities. • Consider requesting increased law enforcement surveillance, particularly of critical assets and otherwise unprotected areas.
	Protection	• Evaluate the need to staff water treatment/production facility at all times. • Consider the need to prohibit recreational use of surface water reservoirs. • Increase security patrol activity to the maximum level sustainable and ensure tight security in the vicinity of mission-critical facilities. Vary the timing of security patrols. • Request that employees change password on critical information management systems.

Source: Kanwal Oberoi, Charleston Water System, Charleston, S.C.

Table 4.2-9 Security Response Procedures for a Severe Threat Level

Condition	Standard Procedures (plus those at lower threat levels)	
SEVERE **(RED)** **Severe Risk of Terrorist Attacks** Signifies a severe risk of terrorist attacks. Protective measures should focus on the decision to close specific facilities and the redirection of staff resources to critical operations.	**Detection**	• Ensure that list of sensitive populations (e.g., hospitals, nursing homes, daycare centers, schools, etc.) within the service area is accurate and shared with appropriate public health officials. • Reconfirm that county and state health officials are on high alert and will inform water utilities of any potential waterborne illnesses.
	Preparedness	• Post daily notices to staff regarding threat level and appropriate security practices. • Where appropriate, place backup operational capacity online (water treatment plan filters, turbines, etc.) • Ensure that key utility personnel are on duty. • Where appropriate, provide public notification for citizens to store emergency water supply or to implement other preparatory measures. • Evaluate the need for opening emergency operations center.
	Prevention	• As appropriate, request increased law enforcement and/or security agency surveillance, particularly of critical assets and otherwise unprotected areas (e.g., consider if National Guard assistance is needed and make appropriate request).
	Protection	• Ensure that existing security policies, procedures, and equipment are effectively implemented. • Re-check security of all on-site chemical storage and utilization areas. • Implement frequent and staggered inspections of the exterior of buildings (to include roof areas) and parking areas. • Re-check the security of critical information systems (e.g., SCADA, Internet, email, etc.) and have staff change computer passwords. • Consider placing staff at remote (typically unmanned) facilities.

Source: Kanwal Oberoi, Charleston Water System, Charleston, S.C.

Example 2

Many water treatment plants undergo construction projects from time to time. When construction is under way, there is an added security burden associated with having contractor personnel coming and going from the plant site. An example security plan for managing contractor personnel is presented in Example 4.2-13.

1.0 PURPOSE

1.1 To provide security instructions for contractors working at the water treatment facility.

2.0 SCOPE

2.1 All utility employees managing contractors working at the water treatment facility.

3.0 RESPONSIBILITIES

3.1 Employees—effectively communicate the security provisions outlined in this work instruction to all contractors you manage at the water treatment facility and communicate all necessary off-access to the contract security personnel and operations personnel.

3.2 Contract Security Personnel—follow all sign-in procedures and provide notification to the proper Utility contact for contractors requesting access to the water treatment facility as outlined in this work instruction.

4.0 WORK INSTRUCTIONS

4.1 Contractors shall enter the facility at the main gate. Access shall not be provided to contractors at the front door to the lab or the pump station gate.

4.2 Security personnel at the main gate shall document sign-in information from the contractor. Sign-in information shall include company name, employee name, date, time, nature of visit, and utility contact person.

4.3 Security personnel shall contact by phone the utility contact person provided by the contractor. Contractors will not be provided access to the facility if contact with designated utility personnel cannot be made.

4.4 The utility contact person shall meet the contractor at an entrance door of one of the buildings within the water treatment facility.

4.5 Prior to exiting the facility, contractor personnel shall sign out with the security personnel.

4.6 When contractors require access to the facility after 5:00 p.m. Monday–Friday or on weekends or holidays, the utility employee who manages the contractors shall contact utility operation personnel and advise them that the contractor will be accessing the facility during off-hours. The contractor shall contact the water treatment control room remotely at the main gate. Utility operation personnel shall initiate sign-in/sign-out procedure or contact the off-duty police officer to meet the contractors at the gate.

4.7 In the event of scheduled or long-term activities requiring the presence of contractors in the facility, the following procedures apply.

- The project manager may provide security personnel a list of the contractor personnel working on site each day.
- The list provided to security personnel from the project manager shall also include location of work, anticipated length of project and utility contact persons.
- The project manager shall report all changes regarding contractor personnel, project dates, project location, or utility contact persons to security personnel immediately.
- Contractor personnel shall sign out with security personnel prior to exiting the facility.

Example 4.2-13 Security Plan for Contractors Working at Treatment Plant (continues)

In the event contractors require access to the facility after 5:00 p.m. Monday–Friday or on weekends or holidays, the project manager shall provide utility operations personnel the above listed information. All sign-in and sign-out procedures shall be followed.

PO Approval: ______________________ Date: ____________

Revision No.	Issue Date	Reason for Revision
1.0	9/08	Issued
2.0	6/10	Reviewed

Policy Update: Utility reserves the right to review and update its policies and procedures as deemed necessary, at any time.

Source: Steve Hubbs.

Example 4.2-13 Security Plan for Contractors Working at Treatment Plant (continued)

4.2.11 Health and Safety Management

The plant shall document health and safety procedures that are designed to safeguard the employees and visitors. The documentation may be specific to the plant or part of a company-wide program.

Rationale

Water treatment plants generally have a number of health and safety programs to protect operating staff and visitors to the plant. These can range from specialized training for confined space entry to making sure that Material Safety Data Sheets (MSDSs) for all chemicals on site are easily available for reference. Documentation of safety procedures can be accomplished in a single safety plan or program description or in several documents that address particular safety subjects. Some documents may pertain only to the plant, and others may address safety issues across the water system.

In any case, the objective of a safety program is to ensure that staff members are safe in the working environment.

Example of Methods or Procedures

Tampa Bay Water has a number of safety documents that apply to water plant operations, including a Hazard Communication Manual and a Personal Protective Equipment Manual. The Hazard Communication Manual contains information about labels and other forms of warning, MSDSs, and employee training and dissemination of hazard information. Example 4.2-14 is the introduction to the Hazard Communication Manual and describes the purpose of the document. Both the Hazard Communication Manual and the Personal Protective Equipment Manual identify appropriate legal references for obtaining additional information.

Hazard Communication Program Purpose and Scope

The Hazard Communication Program is based on a simple concept—that employees have both a need and a right to know the hazards and identities of the chemicals they are exposed to when working. They also need to know what protective measures are available to prevent adverse effects from occurring. The Hazard Communication Program is designed to provide employees with the information they need.

The purpose of this section is to ensure that the hazards of all chemicals produced or imported are evaluated, and that information concerning their hazards is transmitted to employers and employees. The transmittal of information is to be accomplished by means of comprehensive hazard communication programs, which are to include container labeling, other forms of warning, Material Safety Data Sheets (MSDSs), and employee training.

The occupational safety and health standard (29 CFR 1910.1200) is intended to address comprehensively the issue of evaluating the potential hazards of chemicals, and communicating information concerning those hazards and appropriate protective measures to employees, and to preempt any legal requirements of a state, or political subdivision of a state, pertaining to this subject. Under Section 18 of the Occupational Safety and Health Act, no state or political subdivision of a state may adopt or enforce, through any court or agency, any requirement relating to the issue addressed by this Federal standard, except pursuant to a Federally-approved state plan.

Chemical manufacturers or importers are required to assess the hazards of chemicals that they produce or import, and all employers are to provide information to their employees about the hazardous chemicals to which they are exposed, by means of a hazard communication program, labels, and other forms of warning, MSDS, and information and training.

Employers shall ensure that laboratory employees are provided information and training, and laboratory employers that ship hazardous chemicals must ensure that any containers of hazardous chemicals leaving the laboratory are labeled and that an MSDS is provided to other employers.

Employers shall maintain copies of any MSDSs that are received with incoming shipments of sealed containers of hazardous chemicals, shall obtain an MSDS as soon as possible for sealed containers of hazardous chemicals received without an MSDS if an employee requested the MSDS, and shall ensure that the MSDSs are readily accessible during each work shift to employees when they are in their work area(s); and shall ensure that employees are provided with information and training to the extent necessary to protect them in the event of a spill or leak of a hazardous chemical.

This section applies to any chemical that is known to be present in the workplace in such a manner that employees may be exposed under normal conditions of use or in a foreseeable emergency.

Source: Tampa Bay Water, Tampa, Fla.

Example 4.2-14 Purpose and Scope of the Hazard Communication Program at Tampa Bay Water

Questions to Check Progress

1. Is the plant emergency response plan kept up to date with current emergency contacts, resources, and other information that changes frequently?
2. Are all plant staff aware of security plans and their individual role in implementation?
3. Is the plant safety program followed on a daily basis and are visitors included in safety activities?

4.2.12 Customer Inquiry Tracking and Response

If responding to customer inquiries is not the responsibility of plant personnel, this requirement does not apply, except that there must be a system to communicate the results of water quality investigations to plant personnel if a problem is identified. If the plant is responsible for responding to customer inquiries, the plant shall have a system to document and respond to customer inquiries. The documentation may be specific to the plant or part of a company-wide program.

Rationale

Tracking and followup on customer complaints provides a supplementary way for the plant to track water quality and to understand water quality shifts in the distribution system. Customer complaints can be the first indication of a problem for parameters that are not measured daily. For some waterborne disease outbreaks, customer complaints have been the first notification to a plant that there is a treatment problem.

Timely followup on customer complaints is essential to maintaining good customer relations. Customers are often forgiving regarding water quality issues if they are kept well informed about causes and if they have a clear understanding of what they need to do to protect their own health.

Example of Methods or Procedures

Example 4.2-15 shows a customer complaint log sheet from Austin Water. In Austin, customer complaints are directed to the water laboratory, but a similar format could be used by a treatment plant in a smaller system where all the responsibility for water quality falls to the plant staff. Based on the nature of the complaint, water samples are normally taken at the complaint address. Analytical results from the water samples are typically sent to the resident who originally lodged the complaint and then are attached to this form in the tracking system. Information from this tracking system is communicated to the treatment plant for action when the problem is tied to the treatment process, or to distribution system operators for action in the distribution system.

CITY OF AUSTIN
WATER LABORATORY

Circle Sample Type: Complaint, City/Seep, CrossConnection, Metal, Well, Other ______________

Slim Sample ID:______________

Sample Number:__________or copy of Slim Screen attached

Call/Sample taken by/Date/Time: _____/_____/_____

City Water?: ☐YES ☐NO ☐NA

Name:______________	Date Collected/Time: ______________
Sample Address: ______________	Collector: ______________
Name of Complex (if applicable): ______________	
Mailing Address: ______________	Circle Hydrant used: LOH, ROH, LFOH, RFOH, LBOH, RBOH Other: ________
City: ______Zip: ______Grid:______ Map:______	
Phone Number:______________	Site information:______________
Circle Reason: Illness, T/O, Quality, Color, Air, Lead, Taste, Odor, Turbidity Unknown, Other: ________	Physical Description of Sample (smell, color, etc...):
Outside hydrant?☐YES ☐NO ☐ Don't Know	
If YES, Circle Hydrant to be used: LOH, ROH, LFOH, RFOH, LBOH, RBOH, Other: ______	Time sample arrived at lab:______________

Customer Concern (Fully Documented):

(Fully document all Lab communication and action, attach additional pages, e-mails, etc.)

Lab Action (flush, retest, etc...): ______________

Results Reported: How Reported: (Phone, FAX, or Letter)	**To Whom**	**Date/Time**	**Initials (Lab Staff)**	**Comments**

For **City/Seep**, based solely on **this** 1-liter sample and no other criteria, the Water Lab makes the determination that this sample is: __________

Example 4.2-15 Customer Complaint Log Sheet for Documentation of Customer Complaints (continues)

CITY OF AUSTIN
WATER LABORATORY

Circle Sample Type: Complaint, City/Seep, CrossConnection, Metal, Well, Other ____________

CHECKLIST

FOR PROGRAMS

COMPLAINTS, CITY/SEEP, CROSSCONNECTION, METAL, WELL, OTHER

IMPORTANT! All original worksheets are to be attached to the WATER Program.

NOTE: There is a separate Worksheet & Checklist for Complaint samples that are re-collected & re-tested.

ATTACH:	Initials
Complaint Documentation Sheet	
pH Worksheet	
p-Alk/Alk Worksheet	
Hardness/Ca & Mg Worksheet	
Chlorine Residual Worksheet	
T/O Documents	
Turbidity Worksheet	
Micro Bacti Slip	
Ammonia Worksheet	
PO4 Spec Print-out	
Fluoride Worksheet	
Other Special Requests	

Proofed By:_______ Date:____________
QC Proofed By:______ Date:____________

Source: Austin Water, Austin, Texas.

Example 4.2-15 Customer Complaint Log Sheet for Documentation of Customer Complaints (continued)

4.2.13 **Process Reliability**

The treatment plant system shall have sufficient flexibility and equipment redundancy to achieve reliable operation. Examples include

1. Multiple units and equipment for critical treatment processes and systems.
2. An inventory of spare parts, repair parts, and spare replacement equipment on-site sufficient to address routine equipment failures with a minimum of downtime.
3. Provisions to mitigate the impact of less frequent equipment failures that can have serious and immediate effects on the quality of water delivered from the plant. Targeted failures typically involve the disinfectant or coagulant feed systems, with failures occurring at a frequency of once every 3 to 5 years or less. These failures can be highly plant specific or weather related and may be identified through hazard analysis and planning.

Rationale

The critical function of a water treatment plant is to produce potable drinking water, so any process or piece of equipment that could be a single point of failure for the plant should have redundancy or replacement equipment readily available for rapid installation. Plants may struggle to determine what pieces of equipment require redundancy. One way to determine this is to follow the water through the treatment process asking at each piece of equipment or process: "If this equipment or process is down, can the plant still produce potable water?" The resulting list of equipment that cannot be off-line will define the items that must have redundancy or a plan for rapid repair or replacement.

For some pieces of equipment, the entire piece of equipment may not fail, but parts of it could fail easily. For these items, having appropriate spare or replacement parts available in the plant for rapid repair can serve the same function as complete redundancy. Utilities often obtain spare parts for new equipment each time a new piece of equipment is installed.

Example of Methods or Procedures

In designing new facilities or improvements to existing facilities, chemical feed systems are typically provided with two feed pumps to provide redundancy. Figure 4.2-4 shows an example of a process control diagram for a coagulant feed

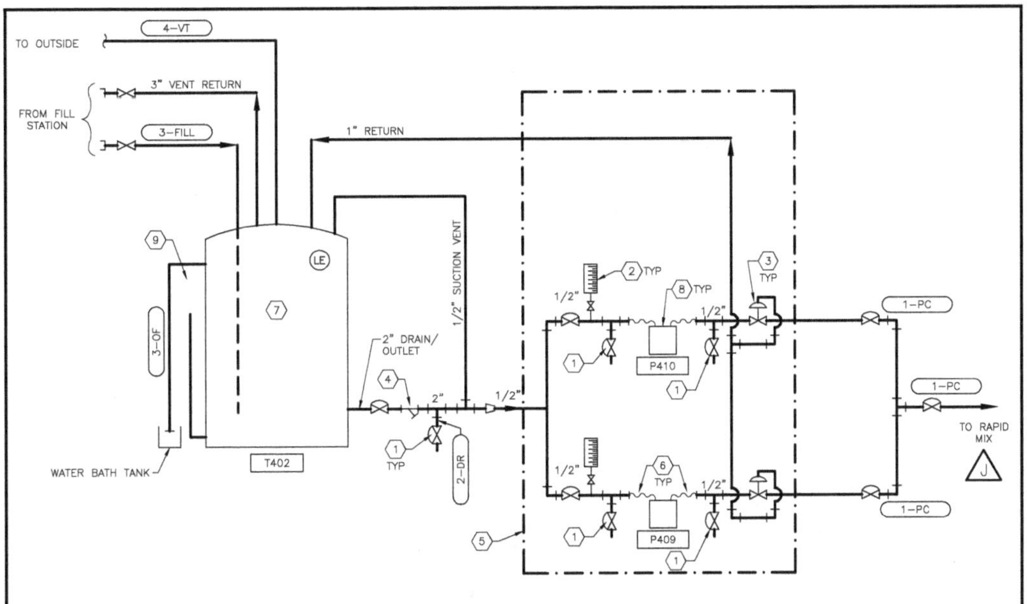

Source: HDR and Jeff Boles, Town of Estes Park, Colo.

Figure 4.2-4 Coagulant Feed System Schematic for the Town of Estes Park

system at the Marys Lake Water Treatment Plant in Estes Park. This system has redundant feed pumps because a failure to feed coagulant will shut the plant down.

4.2.14 **Power**

The plant shall have power redundancy to process water to meet demands and achieve water quality goals. Alternate electrical feeds, standby power generation equipment, or substitute fuels may satisfy this requirement. This requirement may be unnecessary where sufficient storage or alternate production capacity is adequate. Regulatory requirements regarding power reliability must be satisfied.

Rationale

Provision of adequate power redundancy directly affects the ability of the treatment plant to meet the critical function of providing adequate drinking water at all times. Most regulatory agencies have power redundancy requirements that can be met by either provision of an alternative power supply or by adequate stored water to meet demands in the community for a specified period of time. Then if the power supply goes down, adequate treated water can still be provided for a period of time in which it is reasonable to assume that power feed can be restored. Particularly in remote locations where repairs may take a significant amount of time or locations where lightning strikes can easily damage plant power systems, provision of an alternate power feed is necessary.

Questions to Check Progress

1. Do you know where the single points of failure are in your plant?
2. Do you have a rapid repair approach for correcting problems in equipment that are single points of failure?
3. Do you have backup power or adequate system storage to meet the regulations?

4.2.15 Unattended Operation

If unattended operation is permitted, the operation must include the following additional controls:

4.2.15.1 Control limits. Control limits must be set according to the maximum (practical) response time.

4.2.15.2 Alarm notification. The notification protocol for alarms shall be defined for each critical measure. Multiple notifications shall be used when appropriate.

4.2.15.3 Backup notification. Systems must have defined provisions for alternative notification in the event of primary notification system failure (telephone and radio).

4.2.15.4 Default settings. A system failure shall result in automated systems reverting to fail-safe default settings.

4.2.15.5 Notification system validation. The notification system validation program must be documented, and validation tests must be performed quarterly.

4.2.15.6 System failure training. Periodic training on operation during a system failure shall include manual operation of automated systems and situation drills. Training shall include the personnel listed in the notification protocol.

4.2.15.7 SOPs. Plant operation during alarm and failure conditions, including automatic plant shutdown and restart, shall be documented in SOPs.

4.2.15.8 Failure log. The plant shall have a system to retain alarm logs and failure incident records.

Rationale

Plants may be operated in the unattended mode in several different frameworks for operation. Some plants at smaller utilities are always unattended, with a visit from a certified operator once a day. In moderate-sized utilities, plants may have full-time operators on site during the day or for most of the day, but the plant operates unattended during the night. One large utility in Colorado has a 40-mgd water plant that is unattended but tied into a SCADA system and monitored from the system's other large plant. Routine operations are all done remotely, with a stop at the 40-mgd plant once a day by one of the operating staff.

Regardless of the size of the plant and the length of time it is left unattended, all unattended plants must have an electronic system that monitors settings on plant equipment and flow, as well as an automated evaluation protocol and an alarm system that notifies operators of major problems or a plant shutdown. This sort of operational protocol is programmed into a SCADA system that can check control limits, determine when an alarm condition is reached, and initiate an alarm call-out. The electronic system typically keeps a log of alarms and failures.

Example of Methods or Procedures

Example 4.2-16 is a form used to provide standard operating instructions for manual operation of a remote raw water pump station.

Environmental Management System—Standard Operating Instructions

Purpose/Scope: This form is used to prepare Standard Operating Instructions (SOIs) for use by associates who perform operational control and maintenance management tasks where Environmental Management System requirements are in place. The department head (or designee) is responsible for the approval, revision, and issuance of SOIs and ensuring that associates have the necessary training to perform the job.

Instructions: Complete each section below prior to placing new or modified monitoring or control equipment in service. The department head (or designee) shall ensure that maintenance documents are available and controlled at all appropriate locations. NOTE: This form may be replicated on a computer or duplicated on a photocopier. The computer copy must look similar to this document and contain the same information.

Prepared by: Approved by/Date:

Title: Station Remote Manual Mode

1. Purpose: To establish standard operating instructions for Process Control equipment.

2. Scope: To provide standard operating instructions for the Remote Manual Mode for the raw water pump station.

3. Responsibility and Authority: Process Control Shift Supervisor

4. Work Preparation:

4.1 Labor:	4.2 Equipment:	4.3 Materials:

5. Work Step(s):

Safety Requirements
Follow all requirements of the Safety Manual and the applicable safety program or procedure. This may include but is not limited to Lock-Out Tag-Out, Confined Space Entry, Hazard Communications (MSDS), Process Safety Management, Excavation and Shoring, Respiratory Protection, Electrical Safety, or Lab Safety.

Use personal protective equipment predicated by the work environment and material handling requirements. This may include but is not limited to eye protection, face protection, hardhat, foot protection, welder's equipment, life preservers, respirators, gas masks, self-contained breathing apparatus, gas detection equipment, seat belts, hearing protection, or traffic vest. A Safety Manual is located on the intranet under SAFETY heading. MSDSs are located upstairs in Control Room.

Example 4.2-16 Form for Providing Standard Operating Instructions for Remote Manual Operation of a Raw Water Pump Station (continues)

5.1 Operation:

1. Ensure poke point for the Auto-Hand is in the Hand position on SCADA.
2. Ensure pumps are showing available on SCADA.
3. Press poke point on SCADA of desired pump, and press the start command.
4. After pump has started, adjust manual-speed control on SCADA to the desired percentage.
5. Monitor motor windings, bearing temperatures, pump vibration, amps, total dynamic head, and pump submergence.
6. Check for unusual noises and excessive heat coming from the pump.

Shutdown Instructions

1. Ensure poke point for the Auto-Hand is in the Hand position on SCADA.
2. Ensure pumps are showing available on SCADA.
3. Press poke point on SCADA of desired pump, and press the stop command.

In the event of a power loss, place pumps in the off position on SCADA until power is restored. Inform instrumentation of problem and investigate problem. Take corrective action, clear alarms, and start desired pump.

All associates shall be made aware of the consequences of deviating from the task. Failure to follow SOIs may result in additional training and/or disciplinary action at the supervisor's discretion.

The positive significant aspects of following these SOIs would be protection of the water quality, equipment integrity, system integrity, and public safety.

The negative significant aspects from not following this instruction could directly impact water quality, equipment integrity, and contamination of the potable water system, resulting in a threat to public safety for those consuming the water.

6. Related Documents:

- California State University, Sacramento, *Manual of Water Treatment*, Volumes 1 & 2
- Legal and Other Requirements
- Significant Environmental Aspect Ranking
- Standard Operating Instructions
- *Standard Methods for the Examination of Water and Wastewater*

Source: Kanwal Oberoi, Charleston Water System, Charleston, S.C.

Example 4.2-16 Form for Providing Standard Operating Instructions for Remote Manual Operation of a Raw Water Pump Station (continued)

4.2.16 **Human Resources**

Personnel performing work affecting plant operation shall be competent on the basis of appropriate education, training, skills, and experience. The plant shall verify that treatment plant personnel possess valid certification at the necessary level when required by applicable regulations and requirements.

4.2.16.1 Competence, awareness, and training. The plant shall have an education or training program in place to transfer appropriate knowledge, skills, and experience necessary to maintain the competencies and certifications of plant personnel needed to fulfill their assigned tasks. The program must include continued assessment of staff skills and knowledge required to support the mission.

1. Determine the necessary level of competence for personnel performing work affecting the treatment plant.
2. Provide training or take other actions to satisfy these needs.
3. Evaluate the effectiveness of the actions taken.
4. Communicate to its personnel to raise awareness of the relevance and importance of their activities.
5. Maintain appropriate records of education, training, skills, and experience (see Sec. 5.1.3).

Rationale

Operation of today's complex treatment plants requires operators that are well educated and trained to manage a wide range of equipment and understand the interrelationships between multiple treatment goals. Provision of appropriate training is essential to having a well-operated plant. Staffing levels must be appropriate to the plant configuration and size, ensuring adequate labor time to keep the plant operating under all conditions.

A number of methods are available for evaluating the appropriate staffing level for a water treatment plant. Some models specific to the water and wastewater industry evaluate staffing levels by looking at the quantity, type, and complexity of equipment and processes installed at a given treatment facility. Based on similar evaluations of related infrastructure, these models then assign the expected labor requirement in hours.

Work Load Analysis is the preferred method for evaluating labor effort to operate, maintain, and manage water utilities. This method takes into account service level objectives and the unique business environment of the evaluated utility. Work practices such as level of human intervention into process operating events, maintenance practices, reporting requirements, purchasing, laboratory practices, contracted services, and general housekeeping are reviewed to determine how all activities as a whole support the agency's core functions. Based on the aggregate effort and potential labor savings, the present staffing level is either justified or adjusted.

Once a plant has established the number of staff required to operate a plant, evaluation of the core competencies required of those individuals and comparison of those skill requirements to those of current employees provides a way to establish a training program. The Association of Boards of Certification "Need-to-Know Criteria for Water Treatment Operators" (ABC 2007) can be used as a baseline of core competencies to effectively and efficiently operate and maintain water treatment plants. Examples of these target competencies are shown in Table 4.2-10.

Table 4.2-10 Need-to-Know Criteria for Water Treatment Plant Operators and Superintendents

Targeted Core Competency	WTP Superintendent	Senior WTP Operator	WTP Operator
Monitor, evaluate, and adjust water treatment processes	Analyze	Analyze +1	Apply
Operate and maintain equipment	Apply	Analyze	Apply
Perform and develop programs and administrative procedures	Analyze +1	Apply	Comprehend
Comply with drinking water regulations and internal service targets	Analyze +1	Analyze	Apply
Evaluate characteristics of source water	Analyze +1	Analyze	Comprehend
Process control analysis and evaluation (lab analysis)	Analyze	Analyze +1	Comprehend
Supervision/Management/Leadership	Analyze +1	Analyze	Comprehend
Comprehend	Basic level of understanding		
Apply	Understands concepts and can apply to observations. Can submit input and implement corrective action per guidelines.		
Analyze	Recognizes concepts and their relationship. Can diagnose and develop programs to increase effectiveness and efficiency.		

Source: Association of Boards of Certification.

Each of the core competencies listed in the table is comprised of knowledge and ability criteria to further define that competency. For example, the knowledge that is required to monitor, evaluate, and adjust water treatment processes includes the following:

- Chemical application procedures
- General chemistry
- Control systems
- Math concepts
- Physical science
- Treatment concepts and processes

Given this sort of a list, a determination can be made as to what type of training is required for each staff member, and a tailored training program can be developed to meet the specific needs of staff.

Example of Methods or Procedures

Example 4.2-17 shows a list of training resources as identified for a particular utility in Wyoming. This list was the end result after an analysis of the staffing requirements for the utility's two treatment plants and an assessment of the knowledge base of the current operating team. Note that the staff training opportunities range from organized college programs to specialized short training programs offered by AWWA.

Questions to Check Progress

1. If your plant is unattended part of the day, have you documented standard operating procedures?
2. Are staff trained how to run the plant manually?
3. Do you have a regular operator training program?
4. Does your plant evaluate and document the results of operator training?

Training Resources

Traditional community college training provides sound fundamentals. The following community college courses contain basic foundation knowledge:

- Introduction to Chemistry
- Chemistry 1 and 2 and Lab 1 and 2
- Introduction to Organic Chemistry
- PLC (Programmable Logic Controller) Essentials—Basic Course
- PLC Network Systems
- Principles of Network Technology
- Technical Writing
- Fluid Power 1
- Instrumentation Essentials
- Introductory Algebra
- Intermediate Algebra
- Pre-Calculus Algebra
- Introduction to Physics
- Principles of Range Management
- Statistics and the World

Professional Association and Regional Seminars. AWWA Member Section Seminar On-Demand is a program set up by AWWA that provides programmed seminar with instructors for the following:

- Source Water Protection and Planning
- Initial Distribution System Evaluation
- LT2 (Long Term 2 Enhanced Surface Water Treatment Rule) Treatment Options
- Ultraviolet Disinfection
- Security and Vulnerability Assessment

Several committees of the Rocky Mountain Section of AWWA offer regular training on various topics in various locations across the section: the Water Treatment Committee, Action Now Committee, and Distribution System Committee.

Online, Web-based courses by AWWA eLearning—Self-Paced:

- Applied Mathematics
- Basic Mathematics
- Coagulation, Flocculation, and Sedimentation Basics
- Disinfection Basics
- Filtration Basics
- Fundamentals for Chemistry for Water Professionals
- Hydraulics

Please refer to www.awwa.org/Conferences/learning.cfm?ItemNumber=47306&navItemNumber=3415 for course descriptions.

In addition to AWWA, short classes are offered with Continuing Education Units credit from several online professional growth organizations. Http://360water.com offers one-hour short classes on

- Hydraulics
- Fundamentals of Coagulation and Filtration
- Membrane Technology
- Drinking Water Disinfection
- Total Coliform Analysis
- Media Filtration for Drinking Water

Source: HDR Engineering.

Example 4.2-17 Training Opportunities for Water Treatment Plant Staff at a Utility in Wyoming

Sec. 4.3 Plant–Real Property Management and Maintenance

4.3.1 **Maintenance Management**

The plant shall have a maintenance management system with practices adequate to sustain plant performance. The system shall document maintenance frequency goals, and there shall be a system to take corrective action if the goals are not being achieved. The maintenance goals shall include inspection frequency for the process units associated with plant operation.

Rationale

The performance of any water plant is dependent on adequate maintenance of the equipment in the plant. Maintenance management is tied back to reliability by directly impacting the ability of a plant to operate continuously. Maintenance programs are best if they are preventive in approach, rather than only responding to maintenance issues as they occur.

Maintenance programs can fall under one person or several divisions depending on plant size.

Computerized systems are often used by larger plants but are not essential. Maintenance should be performed on all equipment based on the following:

- OEM (original equipment manufacturer) recommendations
- Plant experience
- Predictive maintenance practices such as vibrations analysis, motor circuit evaluation, oil analysis, and thermography
- Calibration where applicable
- Statistical process control programs
- Daily checks by operation staff during inspection rounds

A maintenance program can be combined with an asset management strategy to assist in long-term equipment replacement/upgrade planning and ensure that the plant can run 24/7, avoid equipment obsolescence, and stabilize O&M budget requirements.

Personnel qualified to perform maintenance tasks can be dedicated maintenance staff or may be operators who are cross-trained to complete maintenance tasks. Training for maintenance requires theoretical and practical background, as well as highly developed troubleshooting skills.

Example of Methods or Procedures

An operating plan for water treatment and pumping maintenance is shown in Table 4.3.1. This particular plan ties maintenance goals and accounting for reaching the goals to the utility-wide operating goals. Each quarter, progress toward the goal is logged into this spreadsheet.

Table 4.3.1 Example of Maintenance Goals and Management Program From a Utility Operating Plan

2010 Operating Plan for Process 264: Maintaining Water Treatment & Pumping Maintenance							
Goal Number	**Linkage**	**Action Objective**	**Outcome and Measures**	**Accountability**	**Suppliers**	**Completion Date**	**Results**
1	2007–2021 Strategic Plan Strategy IV (Organizational Effectiveness)	Accomplish Maintenance Skills Training Goals for 2009. Target is 70% of Tier #1 for all individuals. Have two mechanic helpers move to full mechanics.	Outcome: Allows for a more fully developed work force and allows for greater customer response & service. Measures: Actively participate in the instruction & training of the maintenance employees.	Supervisory staff	All staff	Ongoing, measured quarterly	
			5 - Achieved >90% of Tier #1 goals by June 30; Have 2 mechanic helpers (MHs) become maintenance mechanics (MMs) by December 31				
			4 - Achieved >85% of Tier #1 goals by December 31; Have 2 MHs become MMs by December 31				
			3 - Achieved 70% of Tier #1 goals by June 30; Have MH become MM				
			2 - Achieved <70% of Tier #1 goals by December 31				
			1 - Achieved <60% of Tier #1 goals by December 31				
		1st Quarter	Progress is currently at 52.16%.				1.00
		2nd Quarter	Progress is currently at 52.16%. Will update dots during Q3.				1.00
		3rd Quarter	Minimal progress made. Will update training information by the end of the year.				1.00
		4th Quarter	Other training (new equipment), new project/process training was held. This prevented the initial goal from being accomplished.				2.00

Table continues next page

Table 4.3.1 Example of Maintenance Goals and Management Program From a Utility Operating Plan (continued)

Goal Number	Linkage	Action Objective	Outcome and Measures	Accountability	Suppliers	Completion Date	Results
2	2007–2021 Strategic Plan Strategy IV (Organizational Effectiveness)	Demonstrate increased performance focusing on preventive maintenance (PM) work orders. Also track backlog work hours.	Outcome: Allows for increased reliability of filter plant equipment. Measures: Time spent on PM work orders.	Supervisory staff	All staff	Ongoing, measured quarterly	
			5 - PM work orders ≥ 100% for the quarter. All maintenance coordinators (MCs) tracking backlog work hours.				
			4 - PM work orders ≥ 90% for the quarter. All MCs tracking backlog work hours.				
			3 - PM work orders ≥ 80% for the quarter. All MCs tracking backlog work hours.				
			2 - PM work orders ≥ 70% for the quarter. All MCs tracking backlog work hours.				
			1 - PM work orders ≥ 60% for the quarter. All MCs tracking backlog work hours.				
		1st Quarter	PM for the quarter = 77.20%. Backlog hours being tracked. CHFP - ?; DS - 120; Inst - 273.5				2.00
		2nd Quarter	PM for the quarter = 81.26%. Backlog hours being tracked. CHFP - ?; DS - 73; Inst - 243				3.00
		3rd Quarter	PM for the quarter = 87.81%. Backlog hours being tracked. CHFP - 300+; DS - 72; Inst - 188				3.00
		4th Quarter	PM for the quarter = 94.80%. Backlog hours being tracked. CHFP - 380; DS - 362; Inst - 223.5				4.00

Table continues next page

Table 4.3.1 Example of Maintenance Goals and Management Program From a Utility Operating Plan (continued)

Goal Number	Linkage	Action Objective	Outcome and Measures	Accountability	Suppliers	Completion Date	Results
3	2007–2021 Strategic Plan Goals and Objectives, Strategy IV (Organizational Effectiveness)	Provide documentation to ensure system re-certification. Improve utility effectiveness through promoting positive working environment and managing our individual performance. Perform ICS for Operations and Water Quality Departments.	Continue process documentation (Utility Goals and Synopsis Sheet). More cohesive working environment for the good of the department. Measures: Use the composite score from the utility goal results.	Supervisory staff	All staff	Ongoing, measured quarterly	
			5 - Performed two internal surveys for Operations and Water Quality departments during the year. Develop action plan to address gaps and accomplish the action plan.				
			4 - Performed two internal surveys for Operations and Water Quality departments during the year. Develop action plan to address gaps and accomplish some items on the action plan.				
			3 - Performed one internal survey for Operations and Water Quality departments during the year and develop action plan to address gaps.				
			2 - Performed one internal survey for Operations and Water Quality departments during the year.				
			1 - Did not survey.				
		1st Quarter	No surveys issued. Will perform during Q2.				1.00
		2nd Quarter	System Certification paperwork complete. Utility survey performed. Internal customer surveys to be distributed.				2.00
		3rd Quarter	System Certification paperwork complete. Utility survey performed. Internal customer surveys distributed in Nov 2011.				3.00
		4th Quarter	System Certification paperwork complete. Utility survey performed. Internal customer surveys distributed in Nov 2011.				3.00

Table continues next page

Table 4.3.1 Example of Maintenance Goals and Management Program From a Utility Operating Plan (continued)

Goal Number	Linkage	Action Objective	Outcome and Measures	Accountability	Suppliers	Completion Date	Results
4	2007–2021 Strategic Plan Goals and Objectives, Strategy IV (Organizational Effectiveness)	Develop and Implement 3 Process Improvements with the utility. Improvements to be made in the areas of better quality, service, cost control, or efficiency. Formally Submit for Posting on the Intranet. Update quarterly.	Outcome: Implement process & operational improvements. Measure: Completion of process improvements as described.	Supervisory staff	All staff	Ongoing, measured quarterly	
			5 - Achieved greater than 5 Process Improvements with one being an Operational Improvement				
			4 - Achieved 3–5 Process Improvements				
			3 - Achieved 3 Process Improvements				
			2 - Achieved 1–2 Process Improvements				
			1 - Did Not Achieve Goal				
		1st Quarter	Two process improvements in utility discussion.				3.00
		2nd Quarter	Two process improvements in utility discussion.				3.00
		3rd Quarter	Four process improvements in utility discussion. Will write up and submit by end of year.				3.00
		4th Quarter	Achieved 5 process improvements with 2 being operational improvements.				5.00

Table continues next page

Table 4.3.1 Example of Maintenance Goals and Management Program From a Utility Operating Plan (continued)

Goal Number	Linkage	Action Objective	Outcome and Measures	Accountability	Suppliers	Completion Date	Results
5	2007–2021 Strategic Plan Strategy IV (Organizational Effectiveness)	Completion of all PHA, PSM, and Annual Inspection items in each respective work area. PHA's actions completed by 3/19/10; PSM and Annual Inspection Actions completed by 12/31/10.	Outcome: Thoroughly address outstanding issues pertaining to these programs. Measure: Review of listed items.	Supervisory staff	All staff	Ongoing, measured quarterly	
			5 - PHA items completed. PSM and Annual Inspection items completed by 10/31/10.				
			4 - PHA items completed. PSM and Annual Inspection items completed by 11/30/10.				
			3 - PHA items completed. PSM and Annual Inspection items completed by 12/31/10.				
			2 - PHA items completed only.				
			1 - No items completed.				
		1st Quarter	PHA items completed. Other items are in progress.				2.00
		2nd Quarter	PSM and Annual Inspection items in progress.				2.00
		3rd Quarter	All PHA items completed by 10/31/10. One Annual Inspection item is open. PSM items need to be completed by 12/31/10.				3.00
		4th Quarter	All but 4 PSM items completed by end of year. These 4 will be completed by end of Jan 2011. Annual Inspection item addresssed.				3.00

Table continues next page

Table 4.3.1 Example of Maintenance Goals and Management Program From a Utility Operating Plan (continued)

Goal Number	Linkage	Action Objective	Outcome and Measures	Accountability	Suppliers	Completion Date	Results
6	2010 Safety Goals	Meet company safety goals in respect to safety training, OSHA, and PMVA.	Outcome: Safer work environment and company. Measure: Safety training up to date. Performance level meets expectations.	Supervisory staff	All staff	Ongoing, measured quarterly	
			5 - 100% safety training completed by 10/31/10. Maximum 1 OSHA and 1 PMVA.				
			4 - 100% safety training completed by 11/30/10. Maximum 1 OSHA and 1 PMVA.				
			3 - 100% safety training completed by 12/31/10. Maximum 1 OSHA and 1 PMVA.				
			2 - 70% safety training completed. Maximum 1 OSHA and 1 PMVA.				
			1 - No safety training completed.				
		1st Quarter	Some safety training completed for employees.				2.00
		2nd Quarter	Some safety training completed for employees.				2.00
		3rd Quarter	All CBT and SOP safety training completed by 10/31/10.				5.00
		4th Quarter	All CBT and SOP safety training completed by 10/31/10.				5.00

Source: Steve Hubbs.

4.3.2

Plant Cross-Connection Control

The plant shall have documentation of a cross-connection control and backflow prevention program or be part of a cross-connection control program run by another entity. The program will follow the recommendations found in AWWA Manual M14, *Recommended Practice for Backflow Prevention and Cross-Connection Control* (AWWA 2004), or as required by local regulations. The plant will use isolation (plumbing) and containment strategies to protect the water supply and plant employees. The program shall include testing of backflow prevention devices at least annually or as required by local regulations.

Rationale

Backflow is the undesirable reversal of flow of non-potable water or other substances through a cross-connection into a customer's potable water system or the piping of a public water system. There are two types of backflow:

1. Backpressure Backflow—caused by a downstream pressure that exceeds the upstream supply pressure in a public water system or customer's potable water system due to a high pressure source. This could be caused by pumps, boilers, or storage tanks.
2. Backsiphonage Backflow—caused by a negative pressure (i.e., a vacuum or partial vacuum) in a public water system or customer's potable water system. Backsiphonage can occur when there is a water demand that exceeds the normal delivery capacity of the water supply. This could be caused by nearby fire fighting or a break in a water main or service pipe.

Backflows can be reduced or eliminated by segregating potable and non-potable water piping by providing an air gap or by installing a backflow prevention assembly between the two plumbing systems.

Cross-connection control programs are designed to locate the hazardous points where backflow into the potable water system could occur and to prevent the occurrence of backflow into the water system. Cross-connection control programs are mandated by most drinking water regulatory agencies. Administering cross-connection control programs can be time-consuming, but the benefits from a good program to protecting public health are significant.

Example of Methods or Procedures

The cross-connection control program for the City of Austin, Texas, is based on a city ordinance that is implemented by Austin Water. Internal facilities belonging to the utility must follow the same program as is required for industrial facilities throughout the city. Testing of backflow prevention devices in the water treatment plants follows the schedule defined by the ordinance. Article 2 from the ordinance is shown in Example 4.3-1. More information about the Austin program can be found at www.ci.austin.tx.us/water/wwwssd_cc_bpo.htm.

Article 2
Cross Connection Control Program

§ 15-1-11 CROSS CONNECTIONS PROHIBITED.

(A) A person may not:

(1) install or maintain a potable water supply, plumbing fixture, equipment, or construction device that creates a cross-connection, or allows reclaimed, contaminated, or polluted water, mixtures or other substances, or gases, to enter potable water by back siphoning, backpressure, or other means;

(2) connect an auxiliary water supply to the City's public water system or a private plumbing system unless a backflow prevention assembly or air gap is installed as required by this chapter;

(3) use a chemical or substance that may cause pollution or contamination of the public water system without installing a backflow prevention assembly or device as required by this chapter;

(4) connect to the public water system a mechanism or system designed to return reclaimed or used water to the public water system;

(5) connect a reclaimed water system to the City's public water system or to the potable water system of a customer who receives potable water service from the City's public water system; or

(6) connect a vehicle or equipment capable of producing backsiphonage or backpressure without installing a backflow prevention assembly or device as required by this chapter.

(B) A person may not install a backflow prevention assembly in a private plumbing system, fire protection system, process water system, irrigation system, or other water distribution system connected to the City's public water system unless:

(1) the assembly has been designed, manufactured, and tested in accordance with the standards adopted by the American Water Works Association;

(2) the assembly has been tested and listed as an approved backflow prevention assembly by the University of Southern California Foundation for Cross Connection Control and Hydraulic Research; and

(3) the installation complies with the Utility Criteria Manual.

Source: Austin Water, Austin, Texas.

Example 4.3-1 Article 2 From the Cross-Connection Control Ordinance for the City of Austin, Texas

4.3.3 **Disinfection of Water Treatment Facilities**

Plant facilities shall be disinfected in accordance with the requirements of ANSI/AWWA C651 (Disinfecting Water Mains), ANSI/AWWA C652 (Disinfection of Water-Storage Facilities), ANSI/AWWA C653 (Disinfection of Water Treatment Plants), or ANSI/AWWA C654 (Disinfection of Wells).

4.3.3.1 Bacteriological testing. Bacteriological testing shall be completed in accordance with the applicable AWWA standard (ANSI/AWWA C651, ANSI/AWWA C652, ANSI/AWWA C653, or ANSI/AWWA C654) and in conformance with *Standard Methods for the Examination of Water and Wastewater* and using laboratories that are accredited according to local regulatory standards.

Rationale

Disinfection of facilities carrying treated water is necessary to eliminate pathogens that could adversely impact public health. The disinfection of water treatment facilities before they are placed into service is the subject of the referenced ANSI/AWWA standards. The standards can apply to new facilities or to facilities that have been taken off line to be cleaned, inspected, or serviced. Disinfection may also be required in facilities that, under normal operations, continue to demonstrate the presence of coliform bacteria in the plant effluent. The standards describe a number of methods that can be used for disinfecting facilities, each of which must be tailored to a particular application to fit the local facilities.

Bacteriological testing is required after disinfection to demonstrate that the disinfection process has been successful. The standards describe when and where to sample the water after the disinfection process is complete. If a utility is certified by the regulatory agency to run total coliform samples, then a chain of custody form may not be required. For samples sent to a certified laboratory outside the utility, a chain of custody form is necessary for quality control purposes.

Example of Methods or Procedures

Figure 4.3.3 is an example of a Chain of Custody form used by certified laboratories to track and record information regarding water samples. When samples are collected and sent to an outside laboratory, utilities should be sending this type of form to the lab along with the water sample to ensure that a complete record of the history of the sample is recorded.

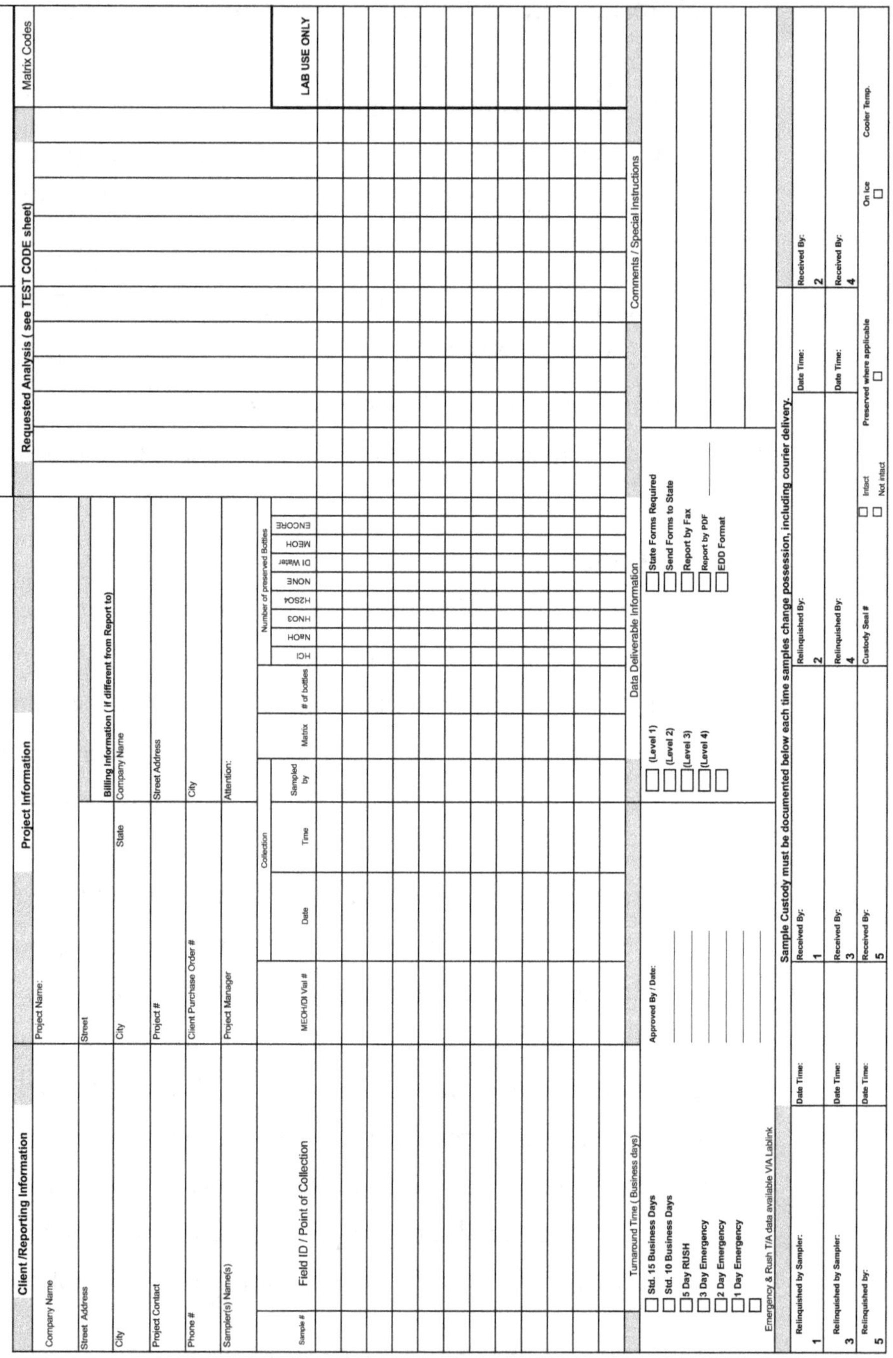

Source: Greg Moore, City of Golden, Colo.

Figure 4.3.3 Chain of Custody Form for Transmitting Water Samples to a Commercial Laboratory From the City of Golden, Colo.

4.3.4 **Operation and Maintenance of Mechanical Equipment**

4.3.4.1 Operating procedures. The treatment plant shall establish written operating and maintenance procedures that document the functioning of every major piece of mechanical equipment for which it is responsible (including emergency power generating equipment). Operating logs or supervisory control and data acquisition (SCADA) shall record operational conditions, such as inlet pressure, discharge pressure, individual run times, flow rate, and other operational variables.

4.3.4.2 Maintenance program. The plant shall have written maintenance procedures for every major piece of mechanical equipment (including emergency power generating equipment), describing frequency, procedures, and maintenance of records. Information shall include basic manufacturer operating requirements. Records shall document inspections and any service performed.

Rationale

Any water plant that has a strong maintenance management system should also have written documentation easily available to maintenance and operating staff which describes the functioning of each major piece of equipment. Operating logs or SCADA data should contain information regarding the run time, pressures, and other variables in operation that each piece of equipment has experienced. By comparing the written description of how the equipment should run and be operated with the data on how that same equipment has been running, the operations staff should be able to determine which pieces of equipment should be in line for service or inspection. By keeping good records of when equipment is serviced and what is done each time, the staff is then able to determine when the next service should occur to maintain all the equipment in good working order. When purchasing new equipment, O&M manuals for each piece of equipment should be obtained from the supplier. If the equipment is part of a complex system, an O&M manual for the system should be required along with equipment delivery.

Example of Methods or Procedures

Example 4.3-2 is a form utilized by Charleston, S.C., for requesting that work be completed on a vehicle. The form provides an organized way to keep a record of evaluation, maintenance, and repair work. A similar format can be used to issue work orders for plant equipment.

Work Request / Courtesy Vehicle / Equipment Visual Inspection ***(To precede all work)***

Veh / Equip #:	**Dept:**		**Contact Person**
Date:	**Time In:**	**Miles / Hours:**	
Work Request / Description:			

To Be Completed by Vehicle / Equipment Technician

Description	OK	Not Ok	Corrected	Deficiencies found & corrected during courtesy inspection
Check all fluid / oil levels				
Radiator condition				
Over flow tank				
Hoses				
Belts				
Battery: cable, water & hold downs				
Air filter				
* Wipers				
* Tires - Wheels - Lugs				
Shocks				
Tail lights				**Comments**
Brake lights				
Turn signals				
Marker lights				
Emergency flashers				
Head lights				
Work lights				
Strobe lights				
* Backup alarm				
* Horn				
* Glass				**Inside & Outside Vehicle Condition**
* Mirrors				
Gauges				
* Brake pedal feel - Operation adjustments				
First aid kit / Eye wash				
* Fire extinguisher - Triangles				
CDL log book				
* Pintle eye hook - Hitch components				
* Steering - Suspension				
* Parking brake - breakaway switch				Tech:
Clutch (Free Travel)				
* Reflectors - Reflector Tape				WO #:
* Exhaust System				
* Air / fluid leaks				

*** May cause D.O.T. - Out of Service**

Corresponding Requirements:	Prepared by:
	Approved by:
Document #:	Approved:

Document Control:

Source: Kanwal Oberoi, Charleston Water System, Charleston, S.C.

Example 4.3-2 Work Request Form for Vehicles From Charleston, S.C.

4.3.5 **Materials in Contact With Water**

Materials that are in contact with the water within the plant (except waste lines and other situations where the water is not part of the production and is not mingled with production water at any point) shall conform to AWWA standards and NSF/ANSI 61 unless the material is not covered by these standards (other appropriate standards may apply in addition to these standards).

Rationale

NSF/ANSI Standard 61: Drinking Water System Components—Health Effects addresses the safety of indirect additives products and materials that come into contact with drinking water or drinking water treatment chemicals. The focus of NSF/ANSI 61, Drinking Water System Components—Health Effects, is the evaluation of contaminants or impurities imparted indirectly to drinking water. By using materials approved under this standard, a utility protects the treated water supply from potential contamination by the equipment that is used to process the water. The products and materials covered include, but are not limited to, process media (e.g., carbon, sand), protective materials (e.g., coatings, linings, liners), joining and sealing materials (e.g., solvent cements, welding materials, gaskets), pipes and related products (e.g., pipes, tanks, fittings), mechanical devices used in treatment/transmission/distribution systems (e.g., valves, chlorinators, separation membranes, point-of-entry drinking water treatment systems), and mechanical plumbing devices (e.g., faucets, end-point control valves).

Manufacturers of equipment used in treating drinking water know and can tell the user which items are approved under NSF/ANSI 61. Specifications for new equipment that will be in contact with drinking water should include a requirement for meeting NSF/ANSI 61.

Questions to Check Progress

1. Does your maintenance staff have access to written O&M information on all pieces of equipment?
2. Are staff trained in maintenance procedures for all types of equipment?
3. Is the plant staff familiar with disinfection methods? Which method has been used most recently in the plant, and is the procedure documented as it was applied locally?

4. Is the operating log reviewed regularly to set preventive maintenance schedules for plant equipment?

Sec. 4.4 Water Quality Management

4.4.1 Performance Goals

In order to protect public health and enhance customer satisfaction, high-quality water must be continuously produced regardless of source water changes or treatment plant component malfunctions.

Drinking water treatment plants shall deliver the quantity of water sufficient to satisfy normal water demands and the quality of water to meet applicable regulatory requirements as well as satisfy the aesthetic demands of customers.

The plant shall adopt a multiple-barrier philosophy that with regard to particulate removal and disinfection, relies on the optimized performance of each major unit process to achieve optimized overall plant performance. The procedures of the Partnership for Safe Water (International Water Treatment Alliance) described in *Self-Assessment Guide for Surface Water Treatment Plant Optimization* (Renner and Hegg 1997) satisfy many of the requirements of this standard. Surface water plants that are participating in this program and have completed the self-assessment should be well prepared to demonstrate conformance with many requirements of this standard: establishing performance goals, monitoring and record keeping comparing actual performance with the goals, and optimization of plant performance on a continuous basis. This program advocates that the plant should be managed to operate efficiently while simultaneously satisfying these objectives. A Partnership for Safe Water Phase III plant in good standing shall be considered to meet the water quality management aspects of this standard as they pertain to particulate removal and disinfection.

4.4.2 Operational Goals and Objectives

The plant shall establish goals for water quality and plant performance, conduct monitoring to determine conformance with the goals, and take appropriate action if goals are not achieved.

4.4.2.1 Goal values. The goals shall provide a margin of safety when compared to applicable regulatory requirements so as to avoid exceeding regulatory limits.

Rationale

Each treatment plant will have performance goals that are generally set at a level which encourages the plant staff to reach for improvement to meet the goals. The treatment plant should set a treatment target for each treatment process so operators will know whether the water is being treated adequately as it moves through the plant. The target for each process might be a single measurement such as turbidity, or multiple parameters, such as a combination of turbidity, pH, and chlorine residual. Targets should be measurable so plant staff can check water quality at key locations throughout the plant. Knowledge of water quality early in the plant treatment train can make a difference in ensuring excellent finished water quality because operators have a chance to make corrections before the water is headed out of the plant.

Example of Methods or Procedures

The example in Table 4.4.2 is a working draft of a plant goal to provide safe and reliable potable water to customers. Because the goal statement is high level, the utility developed strategies to achieve the goal, with associated actions and methods to determine if the strategy is being met. Note that there is quite a difference between this goal statement and its accompanying strategies and a simple numerical treatment goal. This broader approach is more valuable to the utility in the long run because it allows for implementation in steps, back-checking on progress, and revision if needed, while maintaining the overall objective of safe water.

Table 4.4.2 Draft of Plant Goal With Implementation Strategies, Actions, and Measurable Criteria

Goal: Provide Safe and Reliable Potable Water		
Strategy	**Action**	**Measurable Criteria**
Comply with all drinking water regulations.	Continue high-quality treatment. Evaluate disinfection by-product (DBP) compliance forecast and establish DBP reduction plan, if needed. Perform Long Term 2 Enhanced Surface Water Treatment Rule monitoring, establish treatment bin, and install treatment as required.	Achieve 100% compliance for all standards
Use best practices to maintain high-quality treated water.	Jar test regularly. Evaluate filter media on a regular basis. Investigate problem filters when they become apparent.	WTP: Streaming current detector (monitor): 0± 0.5 Zeta Potential: 0 to –5 Settled Turbidity: <1 ntu 90% of time sampling every 15 minutes Settled Turbidity: <2 ntu at all times Start-up Filter Turbidity: ≤0.3 ntu at end of 30 minutes Filter Turbidity: ≤0.1 ntu in 95% of IFE 15 minutes samples; Filter Turbidity: ≤0.15 ntu at all times Filter Particle Count: <25/mL for particles >2 µm in 95% of samples Filter Loading Rate: ≤5 gpm/ft^2 Filter Run Time: >60 hours Filter Rest Before Start-up: ≥3 hours Inactivation: >95% of minimum performance measure Cl_2 Dose: <2.0 mg/L Combined filter effluent Cl_2: 0.4 mg/L

Table continues next page

Table 4.4.2 Draft of Plant Goal With Implementation Strategies, Actions, and Measurable Criteria (continued)

Strategy	Action	Measurable Criteria
Seek alternative methods to increase delivered water quality.	Implement necessary upgrades to the WTP to ensure a multi-barrier system. Improve clarification treatment. Partnership for Safe Water—Phase III Implement AWWA Standard G100-11, Water Treatment Plant Operation and Management. Join Water Research Foundation (WaterRF).	WTP finished water: pH: 7.8 ± 0.2 Alkalinity: ≥40 mg/L $CaCO_3$ Hardness: ≥50 ppm Turbidity: <0.1 ntu Sulfate: <5 ppm Sodium: 5–20 ppm TDS: <125 ppm Conductivity: <00 μmhos/cm Temperature: <20°C Fluoride: 0.9 + 0.1 mg/L Cl_2: 1.1 + 0.1 mg/L Manganese: <0.03 mg/L
	Improve disinfection (install UV).	Meet minimum daily control levels: • 99.99% removal of *Giardia* • 99.999% removal of *Cryptosporidium* • 99.999% removal of enteric viruses
Deliver similar and consistent finished water quality from both plants.	Set uniform levels for water quality parameters. Develop program to bring water from both plants into consistent range for measured parameters. Control pH and/or oxidation reduction potential of finished water.	Maintain finished water: Hardness: >50 mg/L Alkalinity: >40 mg/L pH: 7.8 + 0.2 Turbidity: <0.1 ntu Sulfate: <5 mg/L Sodium: 5–20 mg/L TDS: <125 mg/L Specific Conductance: <200 μmhos/cm Temperature: <20°C No detectable taste and odor Manganese: <0.03 mg/L Fluoride: 0.9 + 0.1 mg/L
Maintain consistent water quality throughout the distribution system.	Minimize effect of mixing regions by implementing consistent multiple-plant finished water quality program. Implement AWWA Standard G200-09, Distribution Systems Operation and Management.	

Table continues next page

Table 4.4.2 Draft of Plant Goal With Implementation Strategies, Actions, and Measurable Criteria (continued)

Strategy	Action	Measurable Criteria
	Develop a monitoring program for the distribution system and a plan for adjusting water quality where necessary.	pH: 7.8 + 0.3 at 95% of locations sampled monthly for total coliforms Free Cl_2: 0.5 + 0.1 mg/L at 95% of locations samples monthly
	Investigate Pb/Cu corrosion behavior and status.	No Lead and Copper Rule violations
	Develop a bacterial regrowth action plan.	Maintain heterotrophic plate count bacteria levels <100 counts/mL in 95% of monthly distribution system samples.
	Develop a plan to meet DBP criteria for reduced monitoring under Stage 1 Disinfectants and Disinfection Byproducts rule (DBPR) and Stage 2 DBPR.	Maintain locational running annual average TTHM (total trihalomethanes) and HAA5 (five haloacetic acids) levels less than 40 μg/L and 30 μg/L, respectively.
	Develop a plan to maintain water quality in distribution system reservoirs.	
Meet all customer demands during normal operations and critical demands during emergencies or droughts.	Acquire adequate water rights.	Maintain • >20 years: No interruption of outdoor use • >100 years: No interruption of outdoor use with major loss • 1,000 years: No interruption of indoor essential use
	Deliver sufficient raw water to treatment plants.	Minimum storage: 3–5 mil gal Minimum source delivery to WTPs: 5 mgd to each plant
Ensure daily reliable plant operations at design flows.	Implement plant maintenance management program.	Minimum production: 10 mgd Maximum production: 16 mgd
Meet daily demand fluctuations and maintain minimum fire flow protection levels.	Provide sufficient storage. Operate to meet demands and to optimize water quality in the distribution system.	Maintain total storage: • Winter Daily: >18 mil gal • Summer Daily: >28 mil gal • Absolute Minimum: 5 mil gal

Table continues next page

Table 4.4.2 Draft of Plant Goal With Implementation Strategies, Actions, and Measurable Criteria (continued)

Strategy	Action	Measurable Criteria
Integrate public health risk factors into source water and treatment management decisions.	Develop a Public Health Protection Index (PHPI) or Risk Index. Establish Stakeholder group to assist in PHPI process Establish consensus of internal and external decision-makers on the PHPI applications.	
Improve knowledge of emerging contaminant occurrence.	Perform the monitoring program included in the Unregulated Contaminant Monitoring Rule. Continue to track unregulated contaminants to determine risk and evaluate monitoring need.	Volunteer for WaterRF participation in research projects of interest.

Source: HDR Engineering.

4.4.3 Monitoring Performance

The plant shall have a monitoring system that can track plant performance and can be used to assess achievement of water quality and performance goals. Monitoring must be in accordance with regulatory requirements. Refer to ANSI/AWWA G400, Utility Management System, for additional details on tracking and measuring improvement.

4.4.3.1 Process supervision goals. The plant shall have specific goals for process supervision, record keeping, process control, procedures for plant operation and upkeep of current procedures, and a system to track operational trends.

4.4.3.2 Sampling and testing goals. The plant shall have specific goals for process control sampling and testing. The type and number of tests shall be defined. Tests shall include regular dosage control evaluations and shall include testing at various control points in the treatment process.

Rationale

Operating a water treatment plant to continuously produce high-quality water requires continuous process monitoring and evaluation. Operators need to know the procedures for checking the treatment process for process control so if any changes occur due to raw water variability or equipment malfunction, adjustments and corrections can be made before the finished water quality suffers. Each plant should set specific locations for regular sampling and at regular intervals around the clock. Diligence in following through with process monitoring can make the difference between excellent finished water and complete process upsets.

Process monitoring can be accomplished with online instruments or grab samples, or by a combination of the two. If online instruments are used, operators must have a regular procedure to review the instrument readings. Analyzed grab samples should be logged in a central location and trends from those results should be documented for continuous review. Most plants have computerized this process either by online readings transmitted directly to a SCADA system that can trend data or by operators entering bench top laboratory results into a computerized tracking system. Either way, the trends are essential tools for understanding the ongoing operation of the plant. Operators should be reviewing trends frequently, looking for changes that may need to be addressed to ensure optimal treatment.

Example of Methods or Procedures

By setting internal plant process goals, one Wyoming water treatment plant has been able to begin optimizing treatment to meet those goals. These goals, shown in Example 4.4-1, allow the plant to adjust internally to strive to meet the goals, and the goals serve as a design target for plant improvements that are planned for the near future.

Plant Treatment Goals

<0.1 ntu finished water goal
<0.5 ntu settled water goal
7.5–7.8 pH finished water target

Taste/Odor Consistency—No Complaints

Filter Run time—72 hr

Water Resource Usage:
- $4,500–$5,000 acre/ft
- Lost water bypassing intake diversion

Chlorine Residual:
- 1.2–1.4 mg/L leaving the plant
- >0.2 mg/L at ends of system

Manage waste streams so they do not affect treatment process.

Source: Wyoming Utility.

Example 4.4-1 Internal Plant Treatment Targets

4.4.4 **Control Measures**

4.4.4.1 Sampling locations. The plant shall select routine water treatment plant sampling locations to verify that the plant is operating properly. The sampling locations shall be reviewed annually and adjustments made based on historical data trends, changes in treatment patterns, or other changes that may affect water quality or plant efficiency. The plant shall analyze data trends and have an action plan to respond to changes. Sample locations shall include, at a minimum, locations required for regulatory compliance monitoring. Additional locations shall be sampled as necessary to provide a complete assessment of water quality in the plant.

4.4.4.2 Sample collection. Samples shall be collected in accordance with *Standard Methods for the Examination of Water and Wastewater* or other applicable procedures.

Rationale

Optimization of water plant operations requires operators to know how each process is functioning at any time. To understand the condition of each treatment process requires collecting appropriate operational data. Sampling locations are key to getting good information about each treatment process. A plan for internal operational monitoring should be well thought out and the result of consultation among utility staff who understand the treatment process and how the individual plant functions. Regulatory requirements as well as process control requirements should be included in the monitoring program. Sometimes sampling is conducted at the same places all the time, regardless of whether that is the best sampling location to meet current needs. While historical data trends can be extremely useful in understanding how a plant operates on a daily basis, sampling points should be evaluated on the basis of the information provided, not just on the basis of the age and quantity of data from a particular site.

Sample collection at each location should be standardized using the appropriate sampling method from *Standard Methods* (APHA et al. 2005). Because multiple operators will be collecting samples at each location over time, all operators should be trained in the sampling procedure and in any analytical procedures that are completed by operating staff.

Example of Methods or Procedures

Figure 4.4.4 shows a schematic diagram of a treatment plant with the regular sampling points identified for process control sampling. Some samples sites are connected to online analyzers and some require operators to take grab samples. Associated with this diagram is a list identifying the frequency of samples at each location.

4.4.5 Corrective Responses

The plant shall have defined corrective responses for each of the water quality goals listed in this standard. The responses must include who shall take action and when to expect correction to achieve the goal.

Rationale

When the treatment processes at a water treatment plant are struggling to maintain adequate treated water quality, plant operators need to know what to do to improve water quality. The treatment goals provide the targets, but a plan of action is necessary to meet those targets, particularly in times of stress. Identifying every possible treatment scenario that may pose a problem is not possible, so operating staff must establish a process for resolving treatment issues. The process may include jar testing or other bench testing that can assist in determining changes in chemical feed. The process may also include requirements to bring in additional help from other operators, technical staff, or managers who can assist in resolving problems.

Example of Methods or Procedures

A standard approach to resolving non-conformities in any area of the utility is illustrated in Example 4.4-2 from Halifax Water. The process can be applied to any non-conforming measurement in the system. It calls for identification of the problem, investigation into causes, and description of actions to correct the problem.

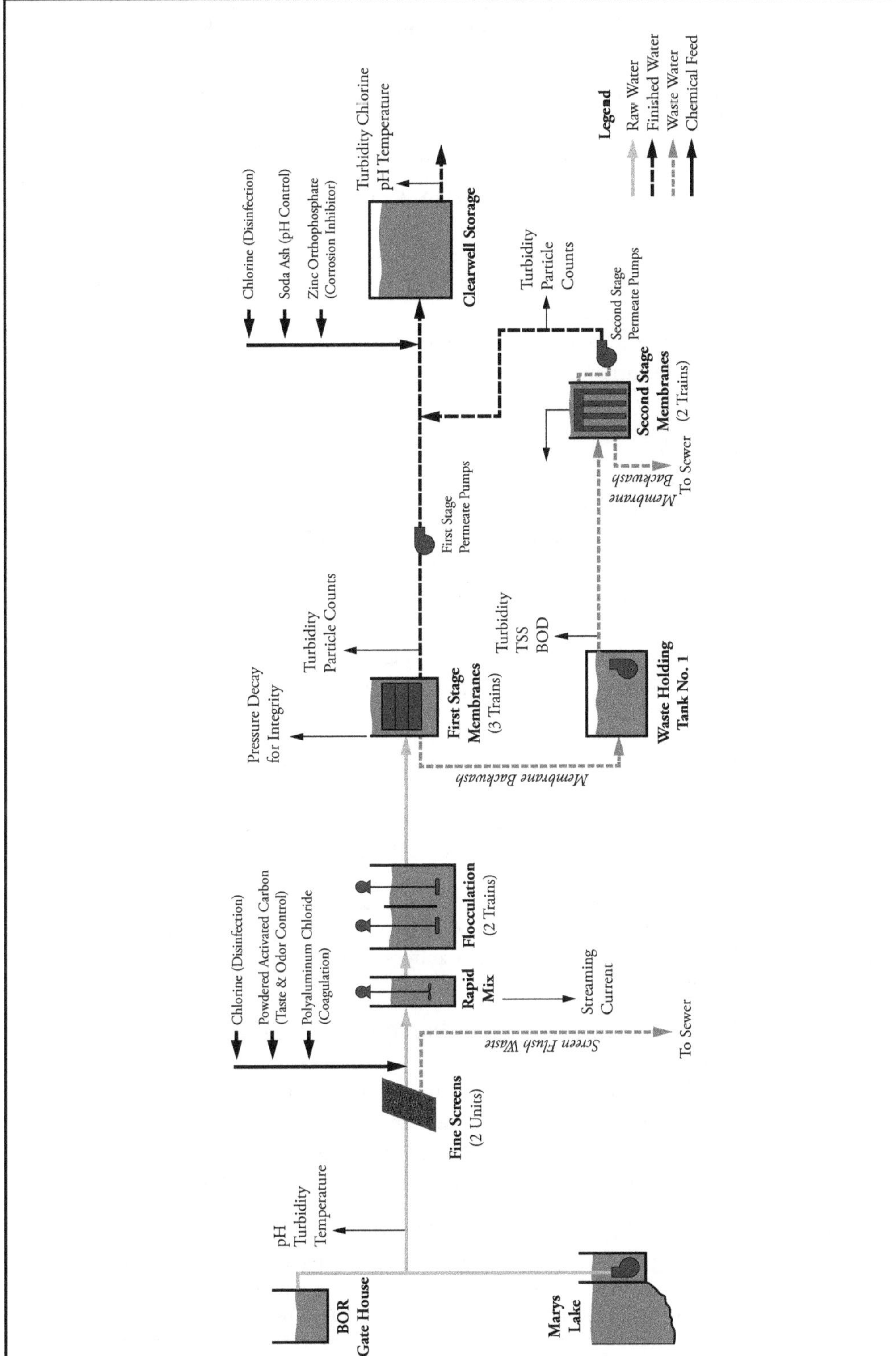

NOTE: TSS = total suspended solids; BOD = biochemical oxygen demand.

Source: Jeff Boles, Town of Estes Park, Colo.

Figure 4.4.4 Process Control Sampling Schematic for the Marys Lake Water Treatment Plant in Estes Park

Halifax Water	Environmental Management System Manual NON-CONFORMITY, CORRECTIVE AND PREVENTIVE ACTION ISO 14001: 4.5.3	Page 1 of 9 HW-EMS-15 March 2011

15 NON-CONFORMITY, CORRECTIVE AND PREVENTIVE ACTION

15.1 Purpose

To describe the method and assign responsibility for:

- Identification of actual non-conformity(ies) and correction and mitigation of their environmental impact;
- Investigation and elimination of the cause(s) of actual non-conformities to prevent their occurrence; and
- Determination of action to eliminate the cause(s) of potential non-conformities to prevent their occurrence.

This includes:

- Actual and potential non-conformities related to the implementation of the Environmental Policy, Environmental Management System (EMS) elements, programs, procedures, and work instructions;
- Achievement of objectives, targets, and legal and other requirements relevant to activities, services, and facilities;
- Accidents and incidents having environmental implications; and
- Findings and observations from internal audits, registration audits, and compliance audits.

Non-conformities can potentially also result from customer complaints, or other inspections (such as Nova Scotia Environment inspections). The Director who receives the complaint or inspection report shall determine if the finding is an EMS non-conformity and, if so, initiate the Non-Conformance–Corrective/Preventive Action process.

15.2 Responsibility

The Director of Water Services is responsible for participating in the investigation of non-conformities and providing resources to support the implementation of corrective and preventive action.

The Superintendent of Water Quality and Treatment is responsible for assigning responsibility and authority for:

- Handling and investigation non-conformities;
- Mitigating any impact caused;
- Designing and implementing corrective action; and
- Reviewing and summarizing actions taken and their effectiveness.

The Plant Supervisor is responsible for participating in the investigation of non-conformities, advising on environmental remedial and mitigative measures, monitoring the effectiveness of remedial and mitigative measures, and recording and reporting the results to the Superintendent of the Water Quality and Treatment and Director Water Services.

Example 4.4-2 Corrective and Preventive Actions for Non-Conformities, Halifax Water (continues)

Halifax Water	Environmental Management System Manual NON-CONFORMITY, CORRECTIVE AND PREVENTIVE ACTION ISO 14001: 4.5.3	Page 1 of 1 HW-EMS-15 September 2011

NON-CONFORMITY, CORRECTIVE AND PREVENTIVE ACTION REPORT
Non-conformance/Observation

Submitted by: Received by: Location: LM___ PW___
Date: Date: Ben___ Gen___

Root Cause Analysis

- Suggested Root Causes:

- Suggested Solutions:

Completed by:

Corrective and Preventive Action:

Action to be taken by:
Action to be completed by (date):
Action approved by:
Date:
Verification of Corrective Action:

Auditor/"Responsible Person":

Date:

Source: Reid Campbell, Halifax Water, Halifax, N.S.

Example 4.4-2 Corrective and Preventive Actions for Non-Conformities, Halifax Water (continued)

Questions to Check Progress

1. Have you set plant goals and operating targets?
2. Does the staff know what the goals are and do they have a stake in meeting the goals?
3. Are the process sampling locations well thought out? Do you have a process for reviewing the sampling locations and revising them based on feedback from operators and the results of the data?
4. Are operational data trends evaluated frequently to determine if a change in treatment is warranted?
5. Do operators know when water quality conditions warrant a change in treatment?
6. Are operators familiar with the action plan for response to water quality changes or treatment issues?

SECTION 5: VERIFICATION

Sec. 5.1 Documentation Required

The plant shall maintain an adequate record-keeping system so that compliance with this standard can be measured. Refer to ANSI/AWWA G400, Utility Management System, for general verification requirements.

5.1.1 General

The documentation shall include

1. Documented statements of quality policy and quality objectives.
2. SOPs.
3. Documented procedures and goals required by this standard.
4. Documents needed by the plant to achieve the effective planning, operation, and control of its processes.
5. Control records required by this standard (see Sec. 5.1.3).

NOTE: Where the term *documented procedure* appears within this standard, it means that the procedure is established, documented, implemented, and maintained.

5.1.2 **Control of Documents**

Documents required for this standard shall be controlled. Records are a special type of document and shall be controlled according to the requirements given in Sec. 5.1.3.

A documented procedure shall be established to define the controls needed to

1. Approve documents for adequacy prior to issue.
2. Review and update as necessary and reapprove documents.
3. Ensure that changes and the current revision status of documents are identified.
4. Ensure that relevant versions of applicable documents are available at points of use.
5. Ensure that documents remain legible and readily identifiable.
6. Ensure that documents of external origin are identified and their distribution controlled.
7. Prevent the unintended use of obsolete documents, and, if such documents are retained for any purpose, apply suitable identification to them.

5.1.3 **Control of Records**

Records shall be established and maintained to provide evidence of conformity to requirements and of the effective operation of this standard. Records shall remain legible, readily identifiable, and retrievable. A documented procedure shall be established to define the controls needed for the identification, storage, protection, retrieval, retention time, and disposition of records.

Rationale

Documentation is a useful way to manage and control processes. Writing standard procedures and work instructions ensures quality and competency by stipulating that everyone performs a process the same way. Documented goals provide a clear road map to take an organization from where it is to where it needs to be. If goals are not documented, it is easy to lose sight of the big picture and flounder on unimportant tasks. Documentation should be regularly approved, reviewed, and revised if necessary and be available at all points of use within the organization. Documentation with a standardized format will be easier to maintain.

Effective document management helps eliminate the use of old, outdated documents. As more utility processes are documented, document and records management become critical. Management of documents and records must be a designated task for a particular person who is responsible for keeping the document library up to date. Standard operating procedures should be developed that cover creation, approval, publication, distribution, revision, archiving, and destruction of electronic and paper documents and records.

Documents should be available to all personnel who need to use them. Procedures should be in place to ensure that documents are retained as long as required, because plant regulatory records must be retained for varying lengths of time, depending on the content. If standard procedures are updated, a procedure for recalling older versions and replacing them with the new one should be in place. Regular updating of standing documents (procedures, sampling plans, etc.) is essential.

Example of Methods or Procedure

Example 5.1.3 from the Charleston (S.C.) Water System is a master list for recording all important documents that are required to be controlled.

Master List of Controlled Documents

Purpose/Scope: To identify and record a master listing of controlled documents.

Instructions: An entry shall be made for each controlled document upon approval and updated thereafter as necessary. Additional entries may be made on a new form or on the Master List of Controlled Documents Continuation Sheet. NOTE: This form may be replicated on a computer or duplicated on a photocopier. The computer copy must look similar to this document and contain the same information.

Group(s): Water Treatment Plant Approved by:

Prepared by/Date: Effective Date:

Reference/ID Number	Title of Document	Type of Document	Approved by/ Date	Document Review and/or Exp. Date	Date Review Completed

Source: Kanwal Oberoi, Charleston Water System, Charleston, S.C.

Example 5.1.3 Controlled Documents

Sec. 5.2 Examples of Documentation

Documentation shall be sufficient to support the applicable requirements in Section 4.

The documentation examples in Table 1 are listed according to the Section 4 requirement number. This list may not be sufficient to fully document conformance with this standard.

Table 1 Documentation Examples

Section Reference Number	Description of Document(s)
4.1	Regulatory compliance records • Discharge permits • Discharge test records
4.2.1	Mission statement
4.2.1.1	Annual budget and cost of operation
4.2.1.2	Capital improvement plan
4.2.1.3	Energy management records (usage trends and costs)
4.2.2.1	Production limits
4.2.2.2	Plant production records (unit process and subunit production records)
4.2.4	Operating permits and licenses
4.2.5.1	Operator qualification policy and training records (also Sec. 4.2.15)
4.2.6	Operational control policy, which might include • Water quality performance goals (plant, unit process, and subunit components) • Monitoring plan and test results • Sample locations—frequency, treatment systems, and subsystems • Disinfectant residual results—maximum, minimum, and average statistics • Disinfection by-product results—maximum, minimum, and averages • Corrosion control (water stability) monitoring results • Color, taste, and odor results • Test results for the parameters with water quality goals • Records of corrective actions taken when goals are not achieved • Storage tank detention time, cleaning and inspection records, mixing results • Meter testing records • Pump station operation and maintenance plan—records • Equipment replacement schedule • Materials specifications • Chemical specifications, analysis results, and screening test results
4.2.6.4	Standard operating procedures (also Sec. 4.2.14.7)
4.2.6.7	Equipment calibration records
4.2.7	Environmental mitigation plan
4.2.8	Emergency response plan
4.2.9	Security program
4.2.10	Health and safety procedures
4.2.11	Customer water quality inquiries and responses
4.2.14	Unattended operations records

Table continues next page

Table 1 Documentation Examples (continued)

Section Reference Number	Description of Document(s)
4.3.1	Equipment maintenance records (preventive, corrective) (also Sec. 4.3.4.2)
4.3.2	Plant cross-connection control program (backflow-prevention device test records)
4.3.3	Plant disinfection procedures (disposal of highly chlorinated water procedures)
4.4	Plant goals and targets
5.1.3	Control of documents procedure

SECTION 6: REFERENCES AND RESOURCES

American Public Health Association (APHA), American Water Works Association (AWWA), and Water Environment Federation (WEF). 2005. *Standard Methods for the Examination of Water and Wastewater.* 21st ed. Washington, D.C.: APHA.

American Water Works Association. 2001. Manual of Water Supply Practices. *M19, Emergency Planning for Water Utilities.* 4th ed. Denver, Colo.: American Water Works Association.

———. 2003a. ANSI/AWWA Standard C653, Disinfection of Water Treatment Plants. Denver, Colo.: American Water Works Association.

———. 2003b. ANSI/AWWA Standard C654, Disinfection of Wells. Denver, Colo.: American Water Works Association.

———. 2003c. *Principles and Practices of Water Supply Operation: Water Treatment.* 3rd ed. Denver, Colo.: American Water Works Association.

———. 2004. Manual of Water Supply Practices. *M14, Recommended Practice for Backflow Prevention and Cross-Connection Control.* 3rd ed. Denver, Colo.: American Water Works Association.

———. 2005. ANSI/AWWA Standard C651, Disinfecting Water Mains. Denver, Colo.: American Water Works Association.

———. 2006. Manual of Water Supply Practices. *M20, Water Chlorination/ Chloramination Practices and Principles.* 2nd ed. Denver, Colo.: American Water Works Association.

———. 2009a. ANSI/AWWA Standard G200, Distribution Systems Operation and Management. Denver, Colo: American Water Works Association.

———. 2009b. ANSI/AWWA Standard G400, Utility Management System. Denver, Colo.: American Water Works Association.

———. 2009c. *Selecting Disinfectants in a Security-Conscious Environment.* Denver, Colo.: American Water Works Association.

———. 2011a. ANSI/AWWA Standard C652, Disinfection of Water-Storage Facilities. Denver, Colo.: American Water Works Association.

———. 2011b. ANSI/AWWA Standard G100, Water Treatment Plant Operation and Management. Denver, Colo.: American Water Works Association.

———. 2011c. *Water Quality and Treatment.* 6th ed. Denver, Colo.: American Water Works Association.

———. Forthcoming. ANSI/AWWA Standard G440, Emergency Preparedness Practices. Denver, Colo.: American Water Works Association.

Association of Boards of Certification (ABC). 2007. ABC Need-to-Know Criteria for Water Treatment Operators. Available: www.abccert.org/pdf_docs/abc2007waterntk2.pdf. Ankeny, Iowa: ABC.

Ginley, J.F., and T.A. Humphrey. 2009. *Operational Guide to AWWA Standard G400: Utility Management System.* Denver, Colo.: American Water Works Association.

International Organization for Standardization (ISO). 2004. *Environmental Management Systems: Requirements With Guidance for Use.* ISO 14001. www.iso.org/iso/home.htm.

McTigue, Nancy E., and James M. Symons, eds. 2010. *The Water Dictionary.* 2nd ed. Denver, Colo.: American Water Works Association.

NSF International and American National Standards Institute. 2005. NSF/ANSI Standard 60, Drinking Water Treatment Chemicals—Health Effects. Ann Arbor, Mich.: NSF International.

———. 2007. NSF/ANSI Standard 61, Drinking Water System Components—Health Effects. Ann Arbor, Mich.: NSF International.

Oberoi, K. 2009. *Operational Guide to AWWA Standard G200: Distribution Systems Operation and Management.* Denver, Colo.: American Water Works Association.

Pizzi, N.G. 2002. *Water Treatment Operator Handbook.* Denver, Colo.: American Water Works Association.

Renner, R.C., and B.C. Hegg. 1997. *Self-Assessment Guide for Surface Water Treatment Plant Optimization.* Awwa Research Foundation Report 90736. Denver, Colo.: American Water Works Association and Water Research Foundation.

SECTION 7: AUDIT CHECKLIST

Checklist Question (AWWA Standard G100)	Yes or No	Remarks & Evidence	Percent Complete
4.1 Compliance With Regulatory Requirements Has the utility identified and documented its legal and other requirements? *Verify that documentation exists and is available to those who need it.* Does the utility comply with all identified legal and other requirements? *Verify examples of compliance, which may include reports routinely submitted to the local regulatory agency.* *If any special regulatory orders or treatment technology orders are in progress, verify compliance as per the special regulatory orders or instructions.*			
4.2 Operational Management Practices			
4.2.1 Business Practices 4.2.1.1 Budget and Financial Management Has the utility acquired adequate resources to operate and maintain the plant as designed? Has the utility established an annual operating budget? Does the utility report cost of operation of plant at least annually? Does the utility incorporate input and recommendations from the plant operating personnel in acquiring plant equipment and other resource acquisition related to the plant operations and maintenance? *Review the annual budget.* *Verify the utility's reporting of variance from the recommended resource allocated to actual expenditures.*			
4.2.1.2 Treatment Plant Capital Improvement Planning Does the utility have a comprehensive plant capital improvement plan? Does the plan take into account future growth and capacity requirements? *Verify that the capital improvement plan addresses the gap analysis and path to vision and missions of the company.*			

Checklist Question (AWWA Standard G100)	Yes or No	Remarks & Evidence	Percent Complete
4.2.1.3 Energy Management Program Does the utility have a comprehensive energy management plan, which includes but is not limited to the following? • Monitoring and optimization of the energy use on regular intervals. • Energy utilization as a major factor in the acquisition of new equipment and in future plant improvements and expansions. *View energy inventory/audit records, procedures that are part of an energy management program, and/or interview key personnel.*			
4.2.2 Treatment Plant Production Limits Has the plant established the capacity of each component of the process train? Are process objectives identified and being met? *Review the written policy and documentation illustrating performance of these objectives.*			
4.2.3 Treatment Plant Adequacy of Treated Supply Does the plant have adequate capacity to ensure that system demands are fulfilled by taking into consideration the system storage capacity? Does the utility identify the system demand and have an established plan of action to meet this demand? If not, does the utility have plans to correct this situation and monitor the progress to ensure this objective? *Review documentation and any plans.*			
4.2.4 Housekeeping Procedures and Cleanliness Does the utility keep all areas of the plant in a safe and orderly fashion? Does the utility conduct regular safety and housekeeping inspections and correct any violations in a timely manner? Does the utility have written procedures to operate and maintain the plant in a manner not to impede the water quality? *Observe conditions and practices and review documentation for inspections and corrective actions if available.*			
4.2.5 Operating Permits and Licenses Does the utility have and maintain current licenses and permits issued by the regulatory agency? *Verify the operating permit of the utility to ensure it is still valid.*			

Checklist Question (AWWA Standard G100)	Yes or No	Remarks & Evidence	Percent Complete
4.2.6 Plant Management *Interview and observe management personnel to ensure that practices and policies are in place in accordance with this standard.*			
4.2.7 Quality Assurance Practices 4.2.7.1 Record Keeping and Data Review Does the utility have a written comprehensive policy and procedures for operational control of the plant? *Review operating manuals and/or instructions/ procedures. Interview operators.* Has the plant identified the type of information to be collected for the proper operations and maintenance of the plant and its frequency? Does the plant review this information routinely and use it to optimize the operation and maintenance of the plant?			
4.2.7.2 Record Review Is there a process in place to collect and implement feedback from the associates regarding the adequacy and accuracy of the data? *Verify the routine review of the data by stakeholders for completeness, accuracy, and adequacy.*			
4.2.7.3 Regulatory Reporting Does the plant review the operational and maintenance data with all stakeholders of the plant to verify that the operational and maintenance control requirements are achieved related to all aspects of the operation, including but not limited to water quality, laboratory operation, and operation and maintenance of mechanical equipment? *Review documentation and ask for examples, meeting minutes, etc.*			

Checklist Question (AWWA Standard G100)	Yes or No	Remarks & Evidence	Percent Complete
4.2.7.4 Standard Operating Procedures Does the plant have a policy and/or procedure that comply with the reporting requirements of the permit to operate, identifying the point of contact? *Verify the protocol of responsibility and reporting of these requirements and ensure these are being met as specified.* *Review and verify that documents submitted for regulatory requirements have adequate approvals and meet the necessary requirements.* Does the utility have standard operating procedures for all the critical operations and maintenance activities? Does the utility routinely review and update these procedures by getting feedback from all of the stakeholders? Do these procedures get updated in a timely manner? Do the standard operating procedures incorporate manufacturers' recommendations and best management practices? *Review SOPs and the process for implementing and updating these procedures.*			
4.2.7.5 Treatment Chemicals Quality Assurance Are procedures in place to ensure that all chemicals added to water meet the NSF/ANSI 60 standard or are approved by the regulatory agency. *Review and verify procedures and documentation of chemicals currently being applied.*			
4.2.7.6 Chemical Quality Do procedures require documentation from suppliers and/or testing to ensure that the chemicals added to the water meet the required standards and quality? *Review procedures, records, and documentation.*			

Checklist Question (AWWA Standard G100)	Yes or No	Remarks & Evidence	Percent Complete
4.2.7.7 Equipment and Instrumentation Calibration Does the utility have a comprehensive program to calibrate all equipment and instruments, which includes frequency of calibration as recommended by the manufacturer or regulatory agency? Does the calibration process follow recommended and authorized industry protocol? Is the chemical feed equipment being calibrated and physical inventories being taken on a monthly basis? *Review procedures and records.*			
4.2.8 Environmental Impact Management 4.2.8.1 Disposal of Discharges From the Plant Does the utility dechlorinate water to be disposed of into natural waters? *Review procedures and records for this operation.*			
4.2.9 Emergency Preparedness and Response Does the utility have comprehensive emergency response plans as described in AWWA Manual M19 and recommended by the regulatory agency? Does the utility revise these plans regularly? Are all plant personnel trained regularly for these plans? *Review procedures and emergency preparedness response plans.* *Review training schedule, records, and meeting minutes regarding emergency response plans.*			
4.2.10 Security Planning and Implementation Does the plant have a comprehensive security plan for the plant and reservoirs under its control? Are all treated water storage facilities covered and protected from contamination or provided with additional treatment? *Tour facilities, review site plans/procedures/drawings, and/or interview key personnel.*			
4.2.11 Health and Safety Management Does the utility have comprehensive health and safety procedures to safeguard their employees and visitors? *Review the safety procedures for their scope and effectiveness.*			

Checklist Question (AWWA Standard G100)	Yes or No	Remarks & Evidence	Percent Complete
4.2.12 Customer Inquiry Tracking and Response Is an inquiry call system in place that identifies and tracks water quality concerns? Does the system record the customer's identification, specific inquiry time, result of investigation, and resolution of the inquiry? Are trends tracked, including the number of inquiries compared to annual goals to reduce number of water quality complaints? *View calls log/system for verification.* Are personnel trained to handle these types of inquiries? *View training records/procedures and/or interview personnel.* Are these inquiries communicated effectively and are records reviewed periodically to identify problem areas of the system? *View records and procedures and/or interview personnel.*			
4.2.13 Process Reliability Does the plant have adequate backup equipment and resources for its operation in case of multiple equipment failures? Does the plant have at least one backup for all critical treatment processes and equipment? *Verify that critical assets have been identified and backup equipment is available or emergency procedures are in place to eliminate downtime.* Are spare parts of all critical processes and equipment adequately available on- or off-site? Has hazard planning been performed where all non-critical and non-routine failures are identified and plans to address them in a timely manner outlined? *Review history of downtime and hazard plans.*			
4.2.14 Power In the case of power outage, are the procedures in place to ensure the continued availability of treated water? *Verify the procedures that plant uses to address this.*			

Checklist Question (AWWA Standard G100)	Yes or No	Remarks & Evidence	Percent Complete
4.2.15 Unattended Operation 4.2.15.1 Control Limits Are control limits established for unattended facilities/operation? *Verify the existence of these controls within written documentation.*			
4.2.15.2 Alarm Notification Have all critical applications been identified and alarm levels established?			
4.2.15.3 Backup Notification What are the backup notification procedures in the case of primary system failure?			
4.2.15.4 Default Settings Are default settings established and is the system designed to revert back to these fail-safe settings in the case of system failure?			
4.2.15.5 Notification System Validation Is the notification system validation program documented and are validation tests performed at least quarterly?			
4.2.15.6 System Failure Training Are drills performed to practice operations during system failure?			
4.2.15.7 SOPs Are SOPs established for operations during system failure and plant shutdown/restart conditions?			
4.2.15.8 Failure Log Are logs maintained to record failure incidents?			
4.2.16 Human Resources Do plant personnel have valid certification as required by the regulatory agency? Are personnel provided necessary training to maintain certifications? *Verify that the plant has adequately trained and competent staff to operate and maintain the plant as required by the regulatory agency.*			

Checklist Question (AWWA Standard G100)	Yes or No	Remarks & Evidence	Percent Complete
4.2.16.1 Competence, Awareness, and Training Has the plant operating personnel acquired adequate licenses as per the operating permit and state requirements? *Review regulatory requirements and compare to documentation.* Does the utility have a documented system to ensure that the operating personnel continues to possess adequate skills, competence, and experience necessary to operate the plant as required by the regulatory agency. *Review training schedules and logs.* Verify the training program of the utility to ensure that it provides adequate knowledge, skills, and experience requirements to maintain the competency and certifications as required by the regulatory agency. Has the utility determined the level of competency (training requirements, licenses, etc.) required for all personnel? Is necessary training provided and are records maintained? *Review training matrices and schedules to verify that training needs are being evaluated and addressed. Review training records.*			
4.3 Plant–Real Property Management and Maintenance			
4.3.1 Maintenance Management Is maintenance performed on a structured and routine basis where required? How are these maintenance tasks assigned, completed, and documented? How is corrective action initiated when required? *Review the maintenance management system (if available) and procedures for the management of maintenance tasks.* Is there a written inspection program for plant process units? Does the program outline frequency (routine, periodic, and comprehensive), procedures, and record keeping? *Review records and procedures for process unit inspections.*			

Checklist Question (AWWA Standard G100)	Yes or No	Remarks & Evidence	Percent Complete
4.3.2 Plant Cross-Connection Control Does the utility have a comprehensive cross-connection and backflow prevention program modeled after AWWA Manual M14? *Review procedures that make up the backflow prevention program.*			
4.3.3 Disinfection of Water Treatment Facilities Does the utility have procedures for the disinfection of treatment facilities prior to placing them online? Does the policy follow ANSI/AWWA Standards C651–C654, and address the disposal of heavily chlorinated water into the environment? *Review the procedures for treatment facilities disinfection.*			
4.3.3.1 Bacteriological Testing Is bacteriological testing completed in accordance with ANSI/AWWA Standards C651–C654? *View procedure for Bac-T testing.*			
4.3.4 Operation and Maintenance of Mechanical Equipment 4.3.4.1 Operating Procedures Does the utility have comprehensive written procedures for the operation of all major plant equipment? Are operational conditions recorded in operation logs or SCADA? *Verify that standard operating procedures exist for large equipment and review operational logs for data availability.*			
4.3.4.2 Maintenance Program Does the utility have comprehensive written procedures for the maintenance of all major plant equipment, including inspection? *Verify that standard operating procedures exist and contain basic manufacturer requirements.*			

Checklist Question (AWWA Standard G100)	Yes or No	Remarks & Evidence	Percent Complete
4.3.5 Materials in Contact With Water Are procedures in place to ensure that all materials in contact with potable water meet the ANSI/AWWA or NSF/ANSI 61 standards or are approved by the regulatory agency. *Review and verify procedures and documentation of chemicals currently being applied.*			
4.4 Water Quality Management			
4.4.1 Performance Goals Does the utility have performance goals to protect the public health and enhance customer satisfaction? Does the utility produce water that always meets or exceeds regulatory requirements in terms of quality and demand? *Verify that the plant has procedures for the continual improvement of the quality of the water and improve productivity and efficiency of its operation. This can be achieved by any program offered by the industry organizations. If a member of the Partnership for Safe Drinking Water Program, review the most recent self-assessment.*			
4.4.2 Operational Goals and Objectives Has the plant established goals for water quality and plant efficiency? Are these goals monitored and what actions are taken if they are not met? *Verify that goals are established and parameters are monitored. Review any actions that have been taken as a result of missed goals.*			
4.4.2.1 Goal Values Do the goals exceed the regulatory requirements, yet are realistic and achievable?			
4.4.3 Monitoring Performance 4.4.3.1 Process Supervision Goals Does the plant's monitoring system track plant performance and can it be used to track water quality goals? *Review the monitoring system and verify that it is in accordance with their regulatory requirements.* *Verify the plant's process supervision goals, record keeping, process control, and procedures for the operation and upkeep of current procedures and system for tracking operational trends.*			

Checklist Question (AWWA Standard G100)	Yes or No	Remarks & Evidence	Percent Complete
4.4.3.2 Sampling and Testing Goals Does the plant have a defined type and number of sampling tests, including regular dosage control evaluations and testing at various control points in the treatment process. *Review sampling plan and records.*			
4.4.4 Control Measures Does the utility have a minimum number of sampling sites to meet or exceed regulatory requirements?			
4.4.4.1 Sampling Locations Are the sampling points reviewed annually to ensure that data trends and changes in the treatment processes are taken into account? *Review the sampling plan and procedures for updating. Compare the sampling sites for the utility with applicable regulations.*			
4.4.4.2 Sample Collection Are samples collected in accordance with *Standard Methods for the Examination of Water and Wastewater*? *Review procedures for sample collection.*			
4.4.5 Corrective Responses Are corrective actions defined for all established water quality goals when they are not being met? Are responsible parties and expectations for timeliness addressed? *Verify that corrective action plans or procedures exist and that they outline who is responsible and when actions are required.*			
5.1 Documentation Required			
5.1.1 General Does the documentation for the elements of Section 4 consist of the following: 1. Quality Policy and Objectives 2. SOPs 3. Documented Procedures 4. Planning Documents 5. Required Records			

Checklist Question (AWWA Standard G100)	Yes or No	Remarks & Evidence	Percent Complete
5.1.2 Control of Documents Does the utility have a standard procedure for the approval of documents prior to use? Is there a procedure in place to approve changes to documents and ensure that the most up-to-date version is available? Are documents of external origin identified and controlled? Does the utility destroy obsolete documents or clearly mark them to avoid unintended use? Has the utility established procedures for the identification, storage, protection, retrieval, retention time, and disposition of documents?			
5.1.3 Control of Records Does the utility have a standard procedure for the approval of records prior to use? Is there a procedure in place to approve changes to records and ensure that the most up-to-date version is available? Are records of external origin identified and controlled? Does the utility destroy obsolete records or clearly mark them to avoid unintended use? Has the utility established procedures for the identification, storage, protection, retrieval, retention time, and disposition of records?			
5.2 Examples of Documentation *Verify the documents for adequacy.*			

APPENDIX A: TOXIC CHEMICAL RELEASE RESPONSE PLAN FOR CHARLESTON, S.C., WATER SYSTEM

Toxic Chemicals at Plant

(Indicate storage capacity and location for each chemical and necessary equipment.)

A. Chlorine

A Chlorine Scrubber is available to clean the air and neutralize chlorine gas in the event of a leak. Chlorine is heavier than air; therefore, the air intake ducts for the scrubber are located close to the floor in each of these rooms. In the event that chlorine gas escapes from these rooms during a leak, a cloud of gas will remain close to the ground and will react with the moisture in the air to form a toxic hypochlorous acid gas. Evacuation procedures will call for anyone located close to this area to have an emergency escape pack (10 or 15 minutes) with which to relocate to a designated rally point. Wind direction and relative humidity are critical factors when leaks occur. The installation of the Chlorine Scrubber has greatly reduced the chance that a chlorine gas leak will impact the environment.

B. Anhydrous Ammonia

Anhydrous ammonia is the opposite of chlorine in that it is a basic solution. A leak of ammonia gas into the air will react with moisture to form ammonia hydroxide, a strong basic gas. Currently the ammonia system has no scrubber and ammonia gas is lighter than air. The cloud of gas associated with an ammonia leak will have a tendency to rise and be carried by wind currents. Considering the current safety equipment, ammonia is the most threatening due to the fact that any leaks will enter the environment surrounding the building. Escape during an ammonia leak has the highest possibilities of placing associates in direct contact with a toxic chemical cloud. Escape breathing packs are vital during this type of evacuation procedure.

Accountability During a Chemical Release

A. Responsibility of Supervisors and Facility Coordinators

Supervisors are directly responsible for the orderly, safe evacuation of all associates. Supervisors in each facility within the plant are responsible for the safe evacuation of each associate in that facility at the time of a reported toxic chemical leak.

Supervisors will support and assist the designated Facility Coordinator in their area with the execution of the Toxic Chemical Release Preparedness and Recovery Plan.

Facility Coordinators are appointed by management for leadership at each normally occupied facility during an emergency. Facility coordinators are usually the ranking supervisor at that facility with an alternate who is usually the second ranking supervisor. The Facility Coordinator is directly responsible for the safe evacuation of all associates and visitors at that facility during an emergency. Practice and thorough knowledge of the plan is mandatory and essential to the success of the Toxic Chemical Release Preparedness and Recovery Plan.

B. Responsibility of the Associates

Associates are responsible for the safety of themselves and fellow associates who may need assistance in the emergency. Associates are also accountable for their knowledge of the Toxic Chemical Release Preparedness and Recovery Plan and what their functions are in this step-by-step procedure. Attendance of training sessions is mandatory, and knowledge of the procedures during an emergency is critical. Accuracy is essential for the safety of everyone involved in an emergency.

C. Responsibility of Management and the Plant Construction Inspector

Management is responsible for the Toxic Chemical Release Preparedness and Recovery Plan. Management is expected to ensure that the plan is implemented properly during an emergency and that Facility Coordinators reach the evacuation rally point safely with all associates accounted for. Management will perform a total head count from associate lists, guest register, and from the Security Guard's logbook to ensure that everyone is safe and out of danger. Management will receive and relay information from the Incident Commander and make a decision concerning further evacuation of the plant site. Management and Facility Coordinators will critique the plan after each use to improve the procedure to meet the needs. Because anhydrous ammonia and chlorine have reporting requirements by the United States Environmental Protection Agency (USEPA) for CERCLA (Comprehensive Environmental Response, Compensation, and Liability Act) and Extremely Hazardous Substances, Management is responsible for implementing the notification procedures outlined in Chapter II, Section 3 of the Toxic Chemical Release Preparedness and Recovery Plan.

The Plant Construction Inspector is responsible for contacting all construction site coordinators (construction foremen) and coordinating their safe evacuation from the plant site until it is safe to return. The Plant Construction Inspector will be responsible for all communication with the construction sites during established work hours. NOTE: After hours, process control will need to direct any contractors on site to the proper rally point using the safest evacuation route.

D. Responsibility of Process Control

Process Control personnel shall respond immediately to a toxic chemical release alarm to assess the situation. Based on the assessment, verification of a chemical release condition, Process Control shall mobilize the Water Treatment Plant (WTP) Emergency Response Team and notify the Fire Department using the 911 system. The front gate guard shall be notified of the situation and secure the gate to prevent entry of unauthorized personnel.

After hours, if the front gate is unmanned, the gate should be manned by an operator to allow entry of authorized personnel. The Director of Water Treatment and Assistant Director of Water Treatment shall be contacted. An accounting of personnel shall be conducted by the senior operator on duty and reported to the Incident Commander. NOTE: After hours, process control will need to direct any contractors on site to the proper rally point using the safest evacuation route.

E. Responsibility of WTP HAZWOPER Technician Volunteers

The Emergency Response Team members present when a toxic chemical release occurs are responsible for mobilizing the Haz-Mat Trailer to a probable Incident Command site and establishing a temporary safe zone. The Incident Commander will determine and set up the safe perimeter around the leak site. The Incident Commander will coordinate the WTP Emergency Response Team and the Fire Department to resolve the situation as soon as possible.

F. Communication

Internal communications within the plant should be conducted using plant walkie-talkies and telephones where applicable.

For External Communications, follow protocol found in referenced Standard Operating Procedures.

Toxic Chemical Release Detection Equipment

A. Chlorine Alarm System

The chlorine alarm system consists of four sensor points, a warning siren, a yellow strobe warning light, and a central control cabinet. Two sensors are located in the chlorine cylinder room and two in the feeder room. (Describe locations.) These remote horns are to warn personnel inside of the building and pump station of a chlorine leak. Associates working elsewhere on the site will hear the warning siren and see a yellow flashing light on the ____ end of the building. The Chlorine Alarm System automatically provides an alarm to the Process Control Room via SCADA and starts the Chlorine Scrubber, which cycles the air in the cylinder and feed rooms through a sodium hydroxide solution to neutralize the chlorine gas. The Toxic Chemical Release Preparedness and Recovery Plan are initiated at the first warning and evacuation will begin.

B. Ammonia Alarm System

The Ammonia Alarm System consists of two sensors, a warning siren (different tone form chlorine siren), a white strobe warning light, and the central control cabinet. (Describe locations.) These remote horns are to warn personnel inside of the building and pump station of an ammonia leak. Associates working elsewhere on the site will hear the warning siren and see a white flashing light on the ______ end of the building. The Ammonia Alarm System automatically provides an alarm to the Process Control Room via supervisory control and data acquisition (SCADA) and activates the ammonia feed room exhaust fan to atmosphere. *There is no Ammonia Scrubber System triggered by the ammonia alarm.* The Toxic Chemical Release Preparedness and Recovery Plan is initiated at the first warning and evacuation will begin.

Toxic Chemical Release Safety Equipment

A. Chlorine Scrubber System

This is a system that is capable of removing leaking chlorine gas from ten (10) cylinders and passing it through a solution of 20 percent sodium hydroxide. The system consists of duct work in the cylinder and feed rooms, a 15,000 gallon tank of caustic soda, a reaction chamber where the caustic will be atomized into the chlorine gas, and blowers and pumps to complete the operation. When a chlorine leak occurs, the alarm system will alert personnel and automatically activate the Chlorine Scrubber System. Escaping chlorine gas will be collected via the ducts and will be sucked into the reaction chamber in the scrubber. The chlorine gas will be met by an atomized sodium hydroxide mist, which will neutralize it. The scrubber remains in operation until the alarm sensors fall back below the alarm point, indicating that the leak has been stopped and the atmosphere is safe.

B. Wind Socks and Their Locations

Wind socks are used by Facility Coordinators and associates to determine the direction and relative strength of the wind. This will be the deciding factor for direction of escape for the associates located close to the vicinity of the chemical leak. These wind socks are orange in color and rotate on a ball-bearing assembly to indicate the direction of the wind. There are four (4) wind socks in the plant and their locations are as follows: (Describe locations.). All associates should make themselves familiar with these devices and their locations.

C. Escape Air Packs

The Laboratory and Administrative Office associates are provided with 10 minute emergency escape packs. They consist of a 4,500 psi compressed air bottle, an air bag to cover the user's head, and a strap to carry the unit over the shoulder.

NOTE: there is no fit test required for this unit. The user merely turns on the air, puts the clear bag with hose attached over the head, and places the strap over the shoulder. Once this is done, the user has approximately 10 minutes of breathing air in which to make an escape to safety.

D. Haz-Mat Trailer

The Haz-Mat Trailer is equipped with International Safety Instruments' self-contained breathing apparatuses with approximately 60 minutes of air supply, Level A chemical suits, a one-ton cylinder capping kit, and various other first aid supplies. Special training is required in order to use this equipment.

E. Two-Way Communications During the Emergency

There are two systems in place for communications in the plant. The first is the standard FM walkie-talkie, which has two channels. Channel #1 is the frequency used by the Maintenance Associates and channel #2 is used by Process Control. During an emergency response to a chemical leak, channel #1 will be used for communication to all associates involved in the evacuation. Channel #2 will be cleared for the Emergency Responders.

The First Responders will be provided two walkie-talkies from Process Control upon arrival.

The second communication method will be by telephone or cell phone.

The 800 MHz radios should be avoided for emergency communications unless they are configured for local communication only. The 800 MHz radios are for full company communications and should be used for normal communications.

F. Site Plan With Escape Routes and Alternate Routes

Overall site plans with escape routes under the various wind direction scenarios are contained in pages _____ of this section.

G. The Evacuation Rally Point

This is a central meeting point for all associates and management and for most cases (except for an east wind) will be the wash pad side of the Maintenance Shop. The rally point for an east wind will be the disaster preparedness parking lot outside the middle gate. Facility Coordinators will determine the evacuation route from the escape scenarios contained in pages ________ of this section. Once the evacuation rally point has been reached, a final head count of associates will be performed. Further evacuation plans will be made at this point based on information gathered concerning the emergency.

Lock Box. Located at front gate. Contains front gate access card, facility keys, MSDS information for chlorine and ammonia, and copies of the Emergency Preparedness and Response Plans. An additional box is located ________________.

Activation of the Emergency Preparedness Plan

A. Phase 1—Detection of a Chlorine or Ammonia Leak

A chlorine or ammonia leak is detected. Sensors in the feed or cylinder rooms trigger the aforementioned siren and warning light. If the leak is chlorine, the scrubber system is automatically activated. Process Control will immediately assess the situation and activate the Toxic Chemical Release Preparedness and Recovery Plan as necessary; see Phase 3. Process Control shall notify the Fire Department using the 911 system. Relay information regarding type of leak and release status; contained in room or not contained. If the leak is contained, the scrubber system should be placed in "manual" operation during the initial assessment.

Process Control Operator shall contact the front gate security guard to secure access to the facility. No traffic shall be allowed onsite unless it is emergency personnel (fire, emergency medical services [EMS], or police), Department of Health and Environmental Control, local emergency planning department officials, or associates called in to assist the First Responders. Note that the utility's Emergency Response Team Members may be called in by the Incident Commander for assistance.

B. Phase 2—Safe Evacuation of Associates, Visitors, and Contractors

1. Facility Coordinators will muster all associates and visitors at the rally point inside of the building. A list of rally points for normally occupied buildings is found in Appendix ______. The Construction Inspector will contact all construction site coordinators that a leak has occurred.

   Facility Coordinators will conduct a head count of associates and visitors. Use the Personnel Evacuation Form to provide a checklist of associates and visitors. Visitors will receive instructions.
2. Visitors will stay with the Facility Coordinator or an assigned associate for the evacuation. Facility Coordinators will then contact Process Control of their attempts to begin evacuation and the route to be taken. The Process Control Shift Supervisor or most Senior Operator will then verify yes or no to the intended evacuation route and advise otherwise if not safe. These communications should be done through telephone or plant walkie-talkie.
3. Facility Coordinators will check the closest visible wind sock for wind direction and relative strength. Associates will be instructed to exit the building facing the wind and which exit route to use. Building rally point will always be the same but site rally point may vary under certain conditions for some facilities within the plant. Facility Coordinators or assigned associate will be responsible for keeping visitors safe at all times during the evacuation. Associates and visitors will then proceed upwind to the Evacuation Rally Point. Incident Commander may assist with plume dispersion information and safe evacuation routes.

   All associates except those in the Control Room shall proceed immediately to the plant rally point. When all associates and visitors safely reach the Evacuation Rally Point, a final head count will be performed by Facility Coordinators and Management. Verification by checklists will be performed, including the visitors log and the gate guard's log. Once accounting is complete, relay personnel accounting to the Incident Commander.

   Control Room Operators should take the necessary action to make the room suitable to shelter-in-place. Further evacuation may be necessary and will be determined by the Incident Commander.

   The Incident Commander may call upon plant associates with HAZWOPER Technician Level certification if necessary to assist with the response. The request should be made directly to plant management or the senior supervisors at the plant.

C. Phase 3—First Responses

During personnel accounting at the rally point, utility's emergency response technician will mobilize the Haz-Mat Trailer and a golf cart to a probable Incident Command site, establish a temporary safe zone, and begin the assessment and containment of the leak. The Incident Commander shall update the Fire Chief or ranking officer upon arrival of the situation and provide two plant walkie-talkies to the Fire Department for use during the emergency. The Incident Commander may request HAZWOPER Technician volunteers to assist the First Responders. This will be done using the proper procedures, equipment, and techniques taught during HAZWOPER Technician certification and training. Teams of response personnel will then work together to isolate the toxic chemical release and/or initiate further evacuations.

D. Maintaining Vital Plant Functions During the Leak

Process Control personnel will communicate with First Responders and assist in any manner as necessary. Process Control personnel not involved with the First Responders will maintain operation of the vital treatment equipment until the situation either becomes threatening or is resolved. All non-vital equipment will be taken out of service until the emergency is over. Their role is to maintain safe, good quality drinking water leaving the plant at the proper pressure throughout the event or until told to evacuate.

E. Damage Control and Assessment After the Leak Is Isolated

1. After the leak is repaired, the Incident Commander will instruct management when it is safe to assess damage to the equipment and initiate repairs.
2. Management will put together a special team to investigate the damage following procedures found in Disaster Recovery Plan and make the repairs.
3. When damage repairs are completed, the First Responders will reactivate the system wearing full response gear and check for leaks.

Once the system is deemed safe and operable by the team, the plant will be returned to normal operations.

A committee will be formed to review toxic release response procedures, revise as needed, ensure that all documents are updated and associates are trained.

Training and Practice Drills

A. Emergency Preparedness Routine Training

Practice drills will be done at random without the prior notification to ensure that the procedure can be performed properly during an actual emergency. Drills will begin when the first alarm is activated. All plant associates and visitors are required to participate in the evacuation drill. Contractors will be asked to participate at least on an awareness or observation level based on work schedules/deadlines, and legal concerns for their projects. They will be notified in the event of an actual alert by the Construction Inspector or Plant Management.

B. Emergency Response Certification and Training

First response to a chemical emergency will be the responsibility of the Emergency Response Team with assistance from the Fire Department. The plant may maintain a limited number of associates with a HAZWOPER Technician Certification, on a volunteer basis. Emergency Response Technicians will receive 8-hour annual refresher training. There will be annual training and exercise drills of the plant's Toxic Chemical Release Evacuation Plan, which may involve some or all of the following responders in the local area and HAZWOPER Technicians:

1. Fire Departments/EMS
2. County Emergency Preparedness
3. Department of Health & Environmental Control (local office)

C. Annual Review

The Emergency Response Plan shall be reviewed annually and updated accordingly.

Hazardous Chemicals

A. List of Hazardous Chemicals and Their Locations

Hazardous Chemicals at the Plant (list the locations):

1. Aluminum Sulfate ______________________________
2. Hydrofluosilicic Acid (Fluoride) ____________________
3. Phosphoric Acid (Orthophosphate) ___________________
4. Potassium Permanganate ________________________
5. Lime Hydrate ________________________________

6. Diesel Fuel ____________________
7. Gasoline ____________________
8. Paint & Solvents—Flammable Storage Locker ____________________
9. Waste Motor Oil ____________________
10. Dry Flocculent Polymer ____________________
11. Sodium Hydroxide—Cl_2 Scrubber ____________________
12. Sodium Chlorite ____________________
13. Coagulant Polymer ____________________
14. Powdered Activated Carbon ____________________
15. Antifreeze ____________________

B. Maximum Quantities on Hand (list maximum quantities)

1. Aluminum Sulfate ____________________
2. Hydrofluosilicic Acid (24%) ____________________
3. Phosphoric Acid (75%) ____________________
4. Potassium Permanganate ____________________
5. Lime Hydrate ____________________
6. Diesel Fuel ____________________
7. Gasoline ____________________
8. Paint & Solvents ____________________
9. Waste Motor Oil ____________________
10. Dry Flocculent Polymer ____________________
11. Sodium Hydroxide (20%) ____________________
12. Sodium Chlorite ____________________
13. Coagulant Polymer ____________________
14. Powdered Activated Carbon ____________________
15. Antifreeze ____________________

C. Associated Health and Safety Hazards

1. Phosphoric Acid (Orthophosphate)—Health Threat
2. Aluminum Sulfate—Minor Health Threat
3. Lime Hydrate—Minor Health Threat
4. Sodium Hydroxide—Health Threat
5. Flocculent Polymer—Minor Health Threat
6. Dry Chlorine—Minor Health Threat
7. Sodium Chlorite—Minor Health Threat

8. Hydrofluorosilicic Acid—Minor Health Threat
9. Diesel Fuel—Explosion/Fire/Environmental
10. Paints & Solvents—Explosion/Fire
11. Gasoline—Explosion/Fire
12. Waste Motor Oil—Minor Health/Environmental
13. Coagulant Polymer—Minor Health Threat
14. Powdered Activated Carbon—Health Threat/Explosion
15. Antifreeze—Minor Health/Environmental

Chemical Release Discovered

For spills or leaks of chemicals other than chlorine, ammonia, diesel fuel, and oil, the person discovering it shall immediately notify their supervisor. The supervisor will follow instructions in Phase 3 above.

If the supervisor discovers the spill or leak, they must designate another associate to follow instructions in Phase 3.

The person discovering the spill or leak must inform their supervisor which chemical has been released and must provide an estimated quantity of the chemical spilled.

For spills or leaks of chlorine, ammonia, diesel fuel, and oil, follow the steps in appropriate plan listed below.

Chlorine, Anhydrous Ammonia, Diesel Fuel, or Oils

After notification, the person discovering the spill should take the following actions:

- If possible, isolate the source of the spill—Do this only if there is no potential for exposure to chemical contamination.
- Refer to MSDS or personal protective equipment (PPE) Requirement List—Found in Control Room.
- Obtain necessary PPE and equipment to contain the spill.
- Until response equipment arrives, utilize on-site methods to contain the spill (shovels, straw, oil absorbent material, sandbags, etc.).
- Estimate the total volume of released chemical and report this to the supervisor.

NOTIFICATION PROCEDURES

Supervisor (or designee) will contact Director of Water Treatment and/or Assistant Director of Water Treatment. Director of Water Treatment or Assistant

Director of Water Treatment will need to determine if chemical is considered Extremely Hazardous and if the total released quantity is above the CERCLA minimum threshold for reporting a release. Refer to table found in Appendix _____. If released quantity is below the threshold limit, no reporting of spill is required. If quantity is above the threshold limit, proceed with notification as follows.

The Director of Water Treatment or Assistant Director of Water Treatment will notify Chief Executive Officer (CEO) of chemical release.

The Director of Water Treatment or Assistant Director of Water Treatment will contact the following agencies *in order* and ask the contact for a mailing address for a written followup notification:

- National Response Center 1-800-424-8802 (24 Hour)
- Department of Health and Environmental Control (DHEC) (list phone number) (24 Hour)
- Fire Department Telephone: 911; Fire Chief: (list number)

Provide the following information to each contact:

- Name, address, & phone number
- Specific location of incident
- Date & time incident occurred or was discovered
- Name of chemical or material released
- Whether chemical is considered extremely hazardous
- Source & cause of release
- Total quantity discharged
- Medium into which material discharged
- Amount spilled in water
- Weather conditions
- Name of the carrier or vessel, truck number (if applicable), or other identifying information
- Number & type of injuries or fatalities
- Whether evacuation has occurred
- Estimation of dollar amount of property damage
- Description of current & future cleanup actions
- Names of other agencies notified or to be notified

As soon as practical after the release and notification, provide a written followup emergency notice to the above agencies. Include the following information:

- Actions taken to respond to and contain the release

- Any known or anticipated acute or chronic health risks associated with the release
- Where appropriate, advice regarding medical attention necessary for exposed individuals

Hazardous Chemical Cleanup/Evacuation Procedures

Since hazardous chemicals are threatening to those who perform cleanup, a qualified Emergency Response Team will be in charge of the Hazard Zone. This team will set up the perimeter and determine if evacuation procedures need to be implemented. If this is the case, the Incident Commander will instruct Plant Management to evacuate. Follow procedures (list location) for evacuation and cleanup of spill.

Critique

Incident Review

After any incident (actual leak or drill) the following activities will take place:

- Safety Committee Chairman or Plant Management fills out Emergency Response Incident Form.
- Appropriate parties as assigned by plant superintendent assess effectiveness of plan.
- Revise plan as necessary and forward to document control representative.

Source: Kanwal Oberoi, Charleston Water System, Charleston, S.C. (adapted).

About the Author

Sarah C. Clark, PE, is a senior project manager and senior professional associate in the Denver office of HDR Engineering. She joined HDR in 2001 having come from the City of Austin Water and Wastewater Utility where she was the water regulatory manager. Clark's broad experience incorporates both drinking water process engineering and extensive involvement in drinking water regulatory development. She served on the US Environmental Protection Agency (USEPA) Technical Work Group that provided technical input to the Regulatory Negotiating Committee for both the Stage 1 and Stage 2 Microbial/Disinfection Byproduct (M/DBP) Rules, representing softening utilities primarily and the sanity of all utilities in general. She continues to serve on American Water Works Association's (AWWA's) DBP Technical Advisory Workgroup (TAW), the Total Coliform Rule TAW, and the Perchlorate TAW, all of which provide technical input to the Water Utility Council and to AWWA negotiations with USEPA on regulatory issues. Clark is the chair of the AWWA Water Treatment Plant Standard Committee and serves on the AWWA Standards Council as well as on several other national and regional AWWA committees.

At HDR, Clark has worked on and/or managed a wide spectrum of projects involving evaluation of water quality issues, plant process optimization and operational challenges, and design and construction of major plant improvements. She serves as the HDR national regulatory expert and is responsible for developing the Safe Drinking Water Act regulatory wall chart (every 2 years) and the Safe Drinking Water Act Newsletter (quarterly).

While working for the City of Austin, Clark provided process engineering and optimization guidance to Austin's three water plants throughout her tenure there, including development and execution of multiple pilot test programs and evaluations of treatment. In addition, Clark has previous experience with other national consulting firms and with the US Army Corps of Engineers as a construction project manager.

Clark has a BA degree in history from the University of California at Santa Barbara (1972) and BS (1983) and MS (1991) degrees in civil engineering from the University of Texas at Austin. She is a member of the Civil, Architectural and Environmental Engineering Academy of Distinguished Alumni at the University of Texas. Clark has one daughter and two cats. She is a square dancer and likes to make quilts.

www.ingramcontent.com/pod-product-compliance
Lightning Source LLC
LaVergne TN
LVHW082351200926
848481LV00012B/3528